Safe Use of Wastewater in Agriculture

Hiroshan Hettiarachchi · Reza Ardakanian
Editors

Safe Use of Wastewater in Agriculture

From Concept to Implementation

UNU-FLORES

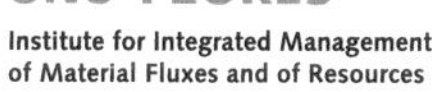

Editors
Hiroshan Hettiarachchi
United Nations University (UNU-FLORES)
Dresden, Sachsen
Germany

Reza Ardakanian
United Nations University (UNU-FLORES)
Dresden, Sachsen
Germany

ISBN 978-3-030-08950-4 ISBN 978-3-319-74268-7 (eBook)
https://doi.org/10.1007/978-3-319-74268-7

Cover image credit: Dr. Olfa Mahjoub

Printed on acid-free paper

This Springer imprint is published by the registered company Springer International Publishing AG part of Springer Nature
The registered company address is: Gewerbestrasse 11, 6330 Cham, Switzerland

Preface

Irrigation is perhaps the most prominent example that we have today to showcase the safe use potential of wastewater. Water scarcity and the cost of energy and fertilizers have driven millions of farmers and other entrepreneurs to make use of wastewater. On a global scale, over 20 million hectares of agricultural land is already irrigated using wastewater. Agriculture, a field that depended on fresh water for irrigation for thousands of years, now needs a more organized and more scientific structure to make use of this opportunity. In order to address the technical, institutional, and policy challenges of safe water reuse, developing countries and countries in transition need clear institutional arrangements and more skilled human resources, with a sound understanding of the opportunities and potential risks of wastewater use.

To address above capacity development needs, a new initiative called "Safe Use of Wastewater in Agriculture" (SUWA) was launched in 2011 by the United Nations Water Decade Programme on Capacity Development (UNW-DPC) led by Reza Ardakanian who is one of the editors of this book. UNW-DPC joined forces with the World Health Organization (WHO), Food and Agriculture Organization (FAO), United Nations Environment Programme (UNEP), International Water Management Institute (IWMI), International Commission on Irrigation and Drainage (ICID), and the United Nations University (UNU-INWEH) to organize a series of SUWA capacity development workshops around the world. Between 2011 and 2013, these capacity development activities brought together 160 representatives from 73 UN member states from Asia, Africa, and Latin America.

The editors of this book represent the United Nations University (UNU-FLORES). As a research institute devoted to promoting sustainability, UNU-FLORES proposes employing an integrated approach to manage water, soil, and waste resources in one nexus to improve the sustainable use of environmental resources. SUWA is an excellent example that fits naturally to this concept—the Nexus Approach. UNU-FLORES has been advocating for SUWA since the inception of its research and capacity development agenda, and has worked with many other UN as well as non-UN organizations, universities, and UN Member States to improve the current understanding and to produce new knowledge.

In 2016, UNU-FLORES received an invitation from the Islamic Republic of Iran, a UN Member State to co-organize a workshop in Teheran to train the country's top officials who are responsible for wastewater management on SUWA basics. The audience in Tehran included representatives from the wastewater management companies from all provinces in the country and other relevant government ministries in addition to the university/research experts. Based on the requests from the co-organizers from Tehran, the workshop focussed more on the implementation aspects of SUWA. Iran as a country has decided to give a prominent place to SUWA in their water/wastewater policy and they wanted to learn more on how a country should commence the SUWA process and overcome the issues related to the policy and implementation aspects. To accomplish this goal, as the UNU-FLORES contribution to the workshop, we made arrangements to identify and bring the best possible team of experts to Tehran who could share their thoughts and prior experience in SUWA implementation issues. In addition to the UNU-FLORES experts, the other contributors represented the KWR Watercycle Research Institute (The Netherlands), International Commission on Irrigation and Drainage (Headquarters in India), German Association for Water, Wastewater and Waste (Germany), Humboldt University Berlin (Germany), National Research Institute for Rural Engineering, Water, and Forestry (Tunisia), and the University of Jordan (Jordan). A former employee of the International Water Management Institute (Headquarters in Sri Lanka) also joined this team as an independent expert.

The expertise of the team and the pool of knowledge that was brought to Tehran were impressive. The event went very well and the contributors received high ratings from the audience as well as local co-organizers. For the benefit of all SUWA stakeholders from around the world, UNU-FLORES wanted to seize this opportunity and make a plan to disseminate this rare combination of knowledge gathered in Tehran. We thought of presenting it as a book edited by UNU-FLORES and chapters authored by the same workshop contributors. What you see in this book is the final result of that thought process. During the writing process, almost all main contributors involved other colleagues from their home institutions, so the final product is in fact a brainchild of well over 15 top experts in the subject area. Like Iran, there are many other countries/regions that are capable of finding/managing the technological aspects of SUWA, but they appreciate learning from others, especially about the policy and implementation challenges and the measures used to overcome them. This book will cater to most of their needs and will provide a very logical and informative starting point.

Editors would like to mention two other recent books that complement the material presented in the current book. The most recent one out of the two is the *Safe Use of Wastewater in Agriculture: Good Practise Examples* edited by H. Hettiarachchi, and R. Ardakanian and published by UNU-FLORES in 2016 (ISBN 9783944863306) and later translated to Spanish, Farsi, and Arabic. This book by UNU-FLORES presents 17 interesting SUWA case studies from around the world. The second book is *Wastewater: Economic Asset in an Urbanizing World,* edited by P. Drechsel, M. Qadir, and D. Wichelns and published by Springer (ISBN 978-94-017-9544-9). This book provides an excellent and detailed coverage of the economic aspects of wastewater. Above two books together with the current book thematically complement each other, with very little but constructive overlap in the subject matter. Anyone interested in implementing SUWA can benefit immensely by referring to all three books, as they collectively present one comprehensive picture of the whole SUWA process.

Dresden, Germany

Hiroshan Hettiarachchi
Reza Ardakanian

The original version of the book frontmatter was revised: Credit line has been included for cover photo. The erratum to the book frontmatter is available at https://doi.org/10.1007/978-3-319-74268-7_9

Contents

Safe Use of Wastewater in Agriculture: The Golden Example of Nexus Approach

Hiroshan Hettiarachchi, Serena Caucci and Reza Ardakanian

Abstract Water, soil, and waste are three key resources associated with agriculture and thus food production as they are closely related to each other. An integrated management of these three resources can bring more benefits to society through increased resource usage efficiency. This approach is commonly known as the Nexus Approach. Safe use of wastewater in agriculture (SUWA) is a simple but powerful example of the Nexus Approach in action. It demonstrates how the sustainable management of one resource in a nexus can benefit the other resources in the same nexus. Wastewater irrigation not only addresses the water demand issues in water stressed areas, but also helps us "recycle" the nutrients in it. The process begins in the waste sector, but the implementation of such a management model can ultimately make a positive impact on the water sector as well as in soil and land management. On a global scale, over 20 million hectares of agricultural land are irrigated using wastewater. Developing countries and countries in transition need clear institutional arrangements and skilled human resources to address the technical, institutional, and policy challenges related to SUWA. From the UN perspective, SUWA also supports achieving some of the key Sustainable Development Goals (SDGs). Taking the wastewater irrigation in the Mezquital Valley in Mexico as an entry point, this chapter builds upon all above facts to provide an introduction to the book and also to illustrate SUWA as a Nexus Approach example.

Keywords Water scarcity · Wastewater · Irrigation · Resources management
Nexus approach · Capacity development

H. Hettiarachchi (✉) · S. Caucci · R. Ardakanian
United Nations University (UNU-FLORES), Dresden, Germany
e-mail: hettiarachchi@unu.edu

H. Hettiarachchi and R. Ardakanian (eds.), *Safe Use of Wastewater in Agriculture*,
https://doi.org/10.1007/978-3-319-74268-7_1

1 Background: Wastewater for Irrigation

Millions of people already live in areas threatened by year-round water scarcity. Mekonnen and Hoekstra (2016) reported that about two-thirds of the global population live under conditions of severe water scarcity at least a month in each year. Geographically, it is interesting to note that the issue is not limited to the countries in the traditionally arid regions such as North Africa and the Middle East. But it also includes some parts of India, China, Central Asia, Sub-Saharan Africa, Central/Western South America, Australia, and North America (WWAP 2016). According to the 2015 World Water Development Report (WWAP 2015) the world is projected to face a 40% global water deficit by 2030, if current conditions continue to persist.

Water scarcity is defined in relation to needs and livelihoods (SEI 2005) and not considered in absolute terms. Often it is about not having sufficient amounts of water irrespective of the quality; water scarcity created by droughts is an example. Communities facing droughts have to find alternatives to mitigate their negative effect on their daily activities and the economy. But quality of water should also be taken into account in determining water scarcity. Pollution may make some water resources that are seemingly available in bulk quantities, not suitable for human consumption. In some areas, the level of contamination in surface water bodies is so high, which makes the water no longer suitable even for non-potable uses such as for agricultural irrigation (FAO 2011; WHO 2016). The agricultural sector accounts for 70–80% of the global human water abstraction and it is estimated to increase by another 70% by 2050 to meet the water demand for over 9 billion people (Lautze et al. 2014). Voß et al. (2012) have theorized that the condition of water scarcity and consequent environmental deterioration can force millions of people to leave their communities and become "environmental refugees" in search of fresh water.

However, we can also find unique examples of other brave communities that did not give up on looking for alternatives because they did not want to become environmental refugees. One such example comes from Mexico. The Mezquital Valley which is located about 160 km north of Mexico City had been an arid area for centuries and by late 1800 they faced a severe shortage of water for irrigation purposes. During the same time Mexico City was also facing a different kind of problem. Mexico City did not have a proper way to dispose its wastewater which was usually collected combined with storm water. The two regions decided to help each other by providing wastewater from Mexico City to the Mezquital Valley for irrigation purposes (Hettiarachchi and Ardakanian 2016b). The slightly lower altitude of the Mezquital area helped them to divert the flow under gravity via a channel and a tunnel (Fig. 1). Through this brave but reckless action the communities found a solution that was "scientific enough" for their time and the agricultural sector began to thrive in Mezquital Valley. No action was taken to look into the safety aspects within the first hundred years. Use of untreated wastewater for irrigation for over a hundred years has of course caused an environmental and public health catastrophes. The high concentration of kidney cancer patients

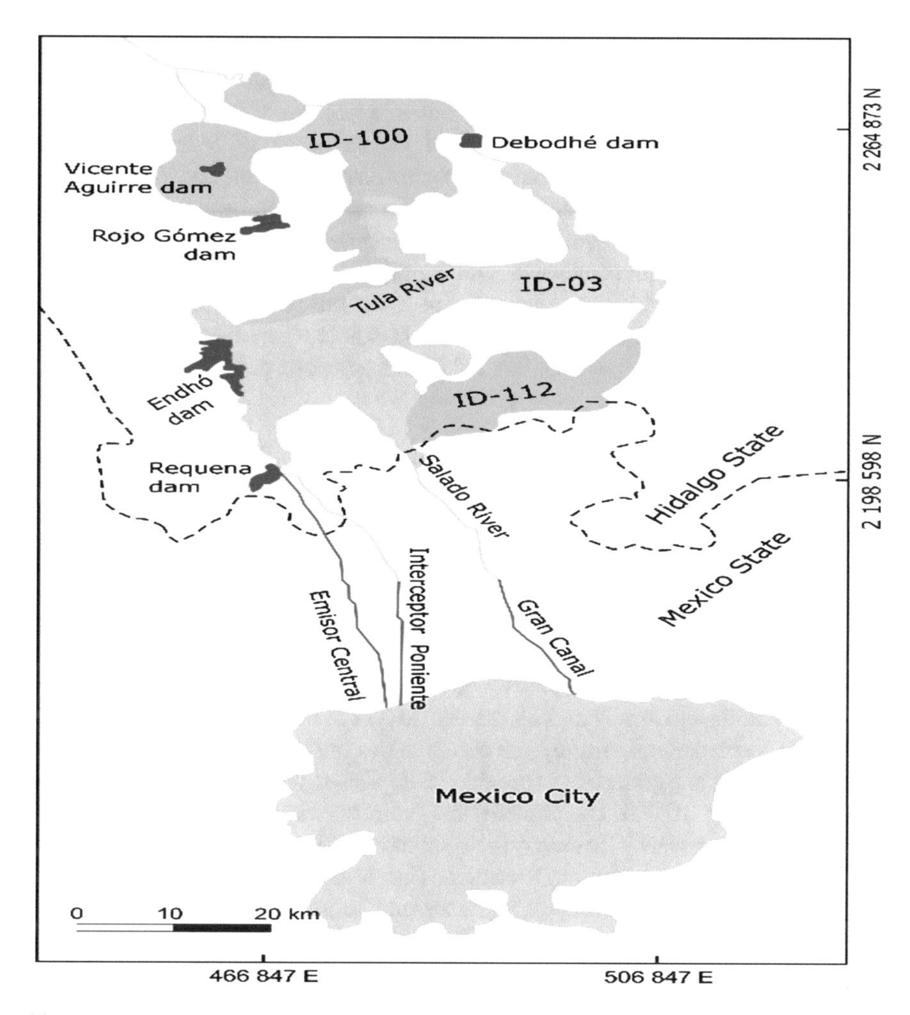

Fig. 1 Location of Mezquital valley, north of Mexico City, and the three irrigation districts [ID: ID-003 (Tula), ID-100 (Alfajayucan), and ID-112 (Ajacuba)] in which untreated wastewater from Mexico City is used (reproduced from Siebe et al. 2016)

reported in this area is believed to be directly related to the irrigation practices using untreated wastewater (Caucci and Hettiarachchi 2017).

The use of wastewater in agriculture makes economic sense, not only in the sense of alleviating water scarcity, but also because wastewater can also fulfil the nutrient requirement for plant growth and thereby eliminate the use of fertilizer. As a consequence, the cost savings from not using fertilizers is translated into a sensible cost reduction within the production cycle. The related improved wastewater management can also lead to the creation of direct and indirect jobs in water-related sectors and beyond (WWAP 2016). The use of wastewater for agronomical purposes can also reduce the stress on freshwater supply and improve the land-use management in arid areas. Thus wastewater reuse in agriculture goes above and beyond the key benefit of being an alternative source of water provisioning.

Mezquital Valley is certainly unique, but it is not alone; there are many other countries and regions that have followed the footsteps of Mezquital Valley, knowingly or not. While some countries such as Israel, Jordan, and Tunisia in the MENA region have shown excellent progress in using treated wastewater together with raised awareness and adjusted policy frameworks, there are many other places where irrigation is still conducted with untreated wastewater. Wastewater is a fundamental part of the water management cycle but so far it has always been considered as a burden instead of a resource. In a time where the need for water is higher than its availability, wastewater is in fact a 'sine qua non' source for water to be considered.

2 Safe Use of Wastewater

The Mezquital Valley example in the previous section teaches us multiple lessons. The overall message is that wastewater irrigation is in fact a viable solution for water shortage issues faced by agricultural communities, if a proper mechanism is put in place to administer the process and assure the safety of the people, crops, livestock, and the environment. The "safety" aspect of the process should be key to its wider acceptance not only by the communities but also by the authorities and institutions. This has given rise to the title "Safe Use of Wastewater in Agriculture" as abbreviated by SUWA. Undoubtedly the "correct" way to practice SUWA is by using treated wastewater. However, there should be a local mechanism to treat and then to distribute the treated wastewater. This also means that there should be acceptable provisions in local policy frameworks to make the process safer.

Paillés Bouchez (2016) provides an interesting account of what went wrong in terms of the safety aspects of water reuse in the Mezquital area. He points out that the "culture of water" in that part of the world did not accept the importance of the treatment of wastewater before reusing. Although practiced in the area for a long time, the reuse concept never became part of the local education, not even at the college level. The awareness started to increase about 20 years ago when the first wastewater treatment plants were introduced. Based on his experience, Paillés Bouchez (2016) also estimates that less than 1% of the people, including teachers and government officers in the communities where the pilot projects of wastewater treatment were implemented, did not know about the safe use aspects. At the same time, thousands of hectares of agricultural land were already being irrigated with untreated wastewater with some kind of participation of national and local governments. Fortunately the safe-use awareness as well as the community support has increased tremendously thanks to the tireless efforts of the Environmental Trust Fund of the State of Hidalgo (FIAVHI). FIAVHI and its associates have implemented more than 80 wastewater management projects in and around Mezquital Valley since 1999; all of them included wastewater treatment targeting reuse applications in agricultural and forestry sectors [unpublished data from Antonio Paillés Bouchez, March 2017].

The story of Mezquital Valley is an example that solidifies the case for establishing institutional arrangements and training programs to ensure the safety aspects in places where wastewater irrigation is practiced. This applies to addressing not only the technical aspects, but also the policy challenges in order to clearly understand the opportunities as well as the potential risks of wastewater use. The "opportunities" are easy to understand and explain as there is a positive and rather quick economic potential. The "risks" are the harder aspects to manage: for one reason, they are not readily seen. Another reason is that the negative economic potential posed by the unmanaged risks is usually not immediate. Even in the case of Mezquital valley, while the economic return is immediate and visible during each cropping season, the impact on the public health component took decades to surface. When the local farmers from the Mezquital Valley were interviewed, the authors got the impression that the key issue is not really about the lack of knowledge on the safety or health issues. The real reason was the ability to postpone thinking about the health issues. For the seemingly minor health issues, they would simply go to a local pharmacy and get some medicine prescribed. These "alternative pharmacies" (farmacia similares) which prescribe antibiotics without any microbiological screening and/or consultation of medical practitioner, seems to be aggravating the situation (Caucci and Hettiarachchi 2017).

Pathogen pollution of surface water is typically high in many countries in Africa, Asia, and Latin America. This obviously translates into a threat of a much higher degree. Communities exposed to microbial polluted water, via agricultural practices or recreational purposes, can be easily subjected to severe disease outbreaks (Hanjra et al. 2012). This is especially true if the source of water contamination is untreated wastewater. If excreta-related disease outbreaks occur, the concentrations of the causative pathogen will increase in the wastewater creating a downward spiral catastrophe (WHO 2006, 2016).

The wastewater use guidelines published by the World Health Organization (WHO) present a framework for the development of national directives and standards for the reduction of health hazards associated with water reuse and also provide information on the monitoring procedures to assure safety. The quality requirements are mainly aimed towards application purposes of the water reuse, and in addition to pathogens also covers salt and nutrient contents for agricultural uses (WHO 2006). Many countries have benefitted from these guidelines. However, literature suggests, over the years, that there is a slight criticism on the adaptability of WHO guidelines on the ground. This is mainly due to the lack of diffusion of the knowledge/regulations by the local authorities to the population. The locals from the Mezquital Valley, who participated in the UNU-FLORES interviews and discussions, were not able to state whether the actual guidelines could be implemented to accomplish SUWA (Caucci and Hettiarachchi 2017). Despite the demonstrated high interest in health aspects, nutrient recovery and economic aspects were dominant in the discussion and the participants could not provide a clear preference on the health issue as a priority in the context of wastewater reuse. More diffusion of knowledge on such guidelines is thus warranted. Wastewater reuse stewardship should be encouraged to demonstrate the fact that the "treated" wastewater can in fact provide nutrients for satisfactory crop yield without compromising health and environmental "safety".

3 SUWA and the Nexus Approach

As a research institute devoted to promoting sustainability, UNU-FLORES proposes employing an integrated approach for water, soil, and waste resources to improve the resource usage efficiency (Hettiarachchi and Ardakanian 2016a). Because of the interrelatedness and interdependencies of these resources, sustainable management of them in an integrated manner can increase resource usage efficiency. SUWA is an excellent example that naturally explains the connection between water, soil, and waste. The process begins in the waste sector, but the selection of the correct management model can make it an asset that is also relevant and important to the water and soil sectors as well. This integrated management model of capitalizing on synergies and tradeoffs is popularly known as the Nexus Approach.

While UNU-FLORES brands SUWA as the "golden example" of the Nexus Approach, the Nexus Approach itself helps in return to maximize the benefits of SUWA. For this reason, UNU-FLORES has been advocating for SUWA since the inception of its research and capacity development agenda, and has worked with many UN as well as non-UN organizations, universities, and UN Member States to improve the current understanding and to produce new knowledge. Especially from the UN perspective the nexus example of SUWA provides ammunition to revisit the current strategies used in water resources management and also another positive discussion point on food security through the topic of nutrient recovery. Therefore the key message here is that SUWA is not just a rudimentary measure of bridging the gap in the volume requirement of water, but it also brings an added benefit of recovery of nutrients for agricultural utilization.

It is true that the inception of SUWA has its roots in the regions severely impacted by water scarcity. However, after years, and now being more organized, SUWA poses the question of why this cannot be applied elsewhere? Irrespective of being in a water scarce area or not, it is interesting to raise the issue of use of freshwater for agricultural, greenspace maintenance, and toilet flushing purposes. Why cannot we push the envelope for sustainability by utilizing this steady and bulk supply of wastewater from a community (especially cities) to cater for above needs of the same community? In many communities, both treated and untreated wastewater is just discharged to the environment all year round or seasonally, thus making it not available for reuse; making it a lost opportunity for production of nutrients or other added values. With the decreasing availability of new sources and increasing demand, water in many regions is becoming too valuable to throw away after "consuming" just once (DWA 2008).

However, who should pay for such treatment and distribution is also a question. There is an argument that the free handing out practiced in some areas in the past had often led to a wasteful handling of the water resource. A future market resulting from wastewater treatment/distribution should stand in a close relationship with the traditional water management options. In this context, the European Water Framework Directive has probably set one good example. The European Water Framework Directive is oriented towards an integrated approach with a cost-covering option for management of water resources (DWA 2008). This type of cycle closing (of water and nutrients) is completely in line with the Nexus Approach. It not only addresses the availability issues, but also reduces the exploitation of fresh water resources.

The second point that was raised early on this section was about the reuse of the nutrients through wastewater irrigation. This is about the vast quantity of nutrients in wastewater that is either lost or misused, adding more weight to the environmental issues under normal circumstances. This has direct relation to food production. The world has not been able to achieve food security even for its current population, but the prediction is that this number will be increased by another two billion within the next 35 years (Hettiarachchi and Ardakanian 2016a). This sends a clear message asking us to look for different potential options. Proper utilization of the nutrients in wastewater through SUWA can very well be one such option.

4 Capacity Development Needs of SUWA

The relatively short history of proper management of wastewater shows us how it can influence society and humanity in a positive way. Confining epidemics is one powerful example. Many believe that the main reason behind the rapid increase in life expectancy during the past hundred years or so is mainly due to the improvement in water quality; undoubtedly proper wastewater management has played an important role in it. Thanks to the continuous dialogues on sustainable themes such as SUWA, now the topic of wastewater management has even transitioned to redefine itself as a tool to establish a circular economy. However, there is a long way to go before we can fully capitalize on the momentum created by these dialogues.

Affordable technologies must be implemented in countries which are lacking them. The choice of technologies is highly site-specific and requires knowledge on climatic systems, levels of economic development, types of economic activity, and level and type of wastewater pollution (UNEP 2015a). Public health has and will always play a central role in the development of wastewater management. The need for having at least secondary-level treatment before reusing highlights how capacity development is urgently needed to support the cause of SUWA. The most appropriate option to reduce the risk in wastewater reuse varies according to the local intended end use and economic factors (O'Neill 2015). Human exposure has to be considered as priority but often despite the fact that *Guidelines for the Safe Use of*

Wastewater, Greywater and Excreta in Agriculture have been developed, their utility for health risk prevention is either very low or not known enough to be applied in the field (WHO 2006).

Safe use of wastewater also requires an active stakeholder participation aiming at the understanding of its benefits and risks (Mahjoub 2013). Unfortunately, poor governance and inadequate attention to operation and maintenance have limited the trust on wastewater sanitation infrastructure by the population and created cultural barriers to sustain habit changes in wastewater reuse in agriculture (Transparency International 2008). Improving wastewater governance therefore requires the understanding of multiple stakeholders' interests which will motivate people in promoting SUWA. Regulatory frameworks need to be tailor-made to the area under implementation with the full respect of the local culture and economy (UNEP 2015a, b).

With urban areas expected to concentrate much of the world's population, growing wastewater volumes are foreseen. Improvement of wastewater management will be fundamental for the achievement of the Sustainable Development Goals (SDGs) set to be achieved by 2030. Specifically, SDG6 targets drastic improvement to sanitation and water quality from where they stand now. In addition, there are several other goals that are closely interrelated to the topic as depicted in Fig. 2 (UN 2015; UN-Water 2016).

Several questions remain. How can we bring the transformational shift of wastewater management into a sustainable process? How can we bring about the recovery of resources in agriculture in a safe manner? Surely the development and adoption of technologies are required but expert knowledge on financing and technical capacity of infrastructures is also among other necessities. In an ideal scenario, the resources should be recovered with no adverse impact to the environment and public health. The process should be cost-effective and supported by policies (UNEP 2015b). The problem in most countries with scarce water resources is not necessarily the lack of rules/regulations on quality standards for water reuse; but rather, above all, a lack of enforcement and monitoring by sufficiently independent, state or public regulatory institutions. The situation makes a strong case for the needed capacity development tools. We believe the content of this book will help us bridge the key gaps from the concept to the implementation of SUWA.

5 What Is Covered in the Book?

The remaining seven chapters of the book present information in three different areas related to SUWA: the justification and the basics technical aspects involved in establishing the concept followed by implementation aspects and finally, the last chapter with a possible alternative to how SUWA is currently being practiced.

After a broader introduction to the opportunities and the risks in SUWA in Chaps. 2 and 3 of the book provides a brief scientific introduction to the Impact of wastewater quality on soils and crops. It focuses on soil and water properties from

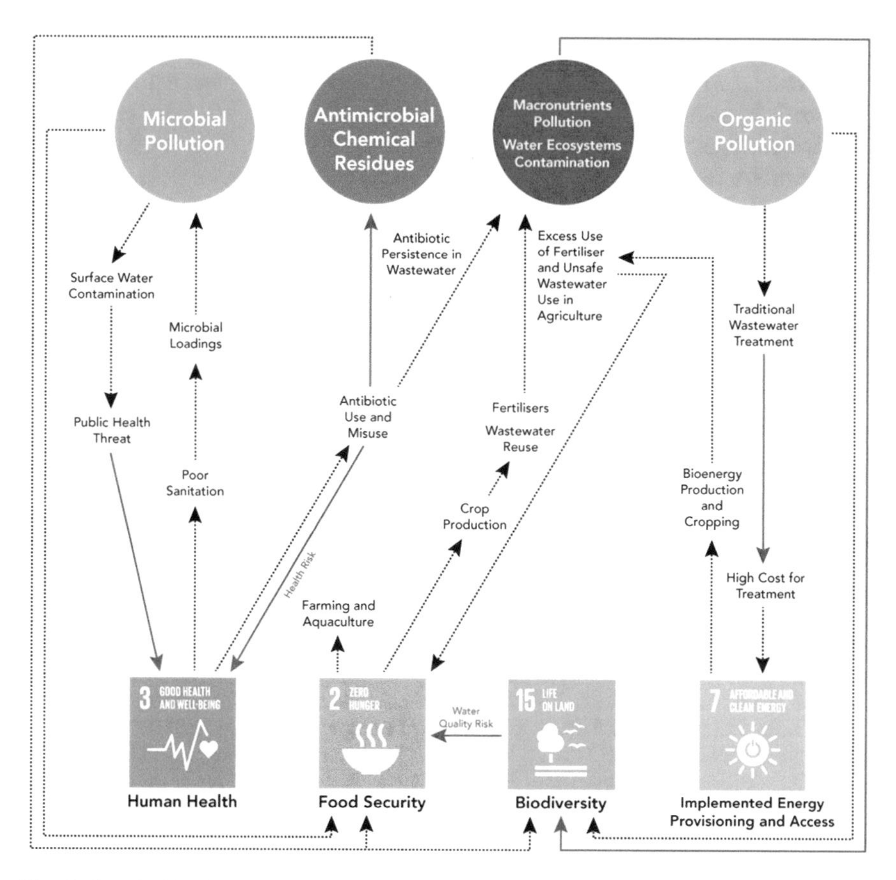

Fig. 2 Sustainable development goals and water quality. Linkages and interactions (modified from UN-Water 2016)

the agricultural point of view and how they may be impacted using wastewater. Wastewater treatment is an essential part of SUWA and the technologies needed for treatment is well established in the published literature. While the technological details of how wastewater treatment should be conducted are not within the scope of the book, it is important to provide some guidance on the selection of appropriate technology for SUWA purposes. This aspect is addressed in Chap. 4 with the help of the German experience, with a matrix developed by the German Association for Water, Wastewater and Waste (DWA).

The rest of the book follows an approach similar to Chap. 4: after presenting information in a general format, the applications are discussed based on a real country-specific experience from different countries. Chapters 5, 6 and 7 cover three important aspects of implementation. Chapter 5 deals with the policy and the governance aspects and more specifically discusses how a country may set up the such framework to become successful with SUWA implementation. In a similar tone, Chap. 6 addresses how a country may set up the framework to improve

SUWA-related health aspects. In both Chaps. 5 and 6, the specific examples are from Jordan, which is a pioneer in wastewater irrigation in the MENA region. Chapter 7 is about public perception/acceptance of SUWA and presents general information together with experiences gained by Tunisia, which is another country in the MENA region with a history of successful SUWA applications.

The final chapter of the book gives some insight into technological advances/alternatives. The chapter specifically focuses on the positive role wastewater can play in managed aquifer recharge (MAR) with the intention of utilizing it later for agricultural purposes. This process can especially have a good influence on public perception and acceptance where direct usage of treated wastewater is not very popular. Appropriately designed and operated MAR with wastewater has proven to be an effective barrier against all pathogens present in wastewater. The specific application examples are from the Netherlands, which has a convincing history of technological breakthroughs in water research.

References

Paillés Bouchez, C.A. (2016). Council for certification of irrigation with treated water in Mexico (Mexico). In H. Hettiarachchi & Ardakanian (Eds.) *Safe use of wastewater in agriculture: Good practice examples* (pp. 279–299). Dresden: United Nations University Institute for Integrated Management of Material Fluxes and of Resources (UNU-FLORES).

Caucci, S., & Hettiarachchi, H. (2017). Wastewater irrigation in the Mezquital Valley, Mexico: Solving a century-old problem with the nexus approach. In *Proceedings of the International Capacity Development Workshop on Sustainable Management Options for Wastewater and Sludge*, March 15–17, 2017, Mexico. Dresden: United Nations University Institute for Integrated Management of Material Fluxes and of Resources (UNU-FLORES).

DWA. (2008). Treatment steps for water reuse. In *DWA-Topics*. Hennef, Germany: Deutsche Vereinigung für Wasserwirtschaft, Abwasser und Abfall e. V. (German Association for Water, Wastewater and Waste, DWA).

FAO. (2011). Agriculture and water quality interventions: A global overview. *SOLAW Background Thematic Report—TR08*. Food and Agricultural Organization (FAO).

Hanjra, M. A., Blackwell, J., Carr, G., Zhang, F., & Jackson, T. M. (2012). Wastewater irrigation and environmental health: Implications for water governance and public policy. *International Journal of Hygiene and Environmental Health, 215*(3), 255–269. https://doi.org/10.1016/j.ijheh.2011.10.003.

Hettiarachchi, H., & Ardakanian, R. (2016a). *Environmental resources management and the nexus approach: Managing water, soil, and waste in the context of global change*. Switzerland: Springer Nature.

Hettiarachchi, H., & Ardakanian, R. (2016b). *Safe use of wastewater in agriculture: Good practice examples*. Dresden, Germany: UNU-FLORES.

Lautze, J., Stander, E., Drechsel, P., Da Silva, A. K., & Keraita, B. (2014). Global experiences in water reuse. In *Resource Recovery and Reuse Series* 4. Colombo: International Water Management Institute (IWMI)/CGIAR Research Program on Water, Land and Ecosystems. www.iwmi.cgiar.org/Publications/wle/rrr/resource_recovery_and_reuse-series_4.pdf.

Mahjoub, O. (2013). "Ateliers de sensibilisation au profit des agriculteurs et des femmes rurales aux risques liés à la réutilisation des eaux usées en agriculture: Application à la région de Oued Souhil, Nabeul, Tunisie" [Awareness-raising workshops for farmers and rural women about the risks related to the use of wastewater in agriculture: Applied to the area of Oued Souhil, Nabeul, Tunisia]. In *UN-Water. Proceedings of the Safe Use of Wastewater in Agriculture, International Wrap-Up Event*, June 26–28, 2013, Tehran. (In French.) www.ais.unwater.org/ais/pluginfile.php/550/mod_page/content/84/Tunisia_Ateliers%20de%20sensibilisation%20au%20profit%20des%20agriculteurs%20et%20des%20femmes%20rurales_Mahjoub.pdf.

Mekonnen, M. M., & Hoekstra, A. Y. (2016). Four billion people facing severe water scarcity. *Science Advances*, *2*(2). https://doi.org/10.1126/sciadv.1500323.

O'Neill, M. (2015). *Ecological sanitation—A logical choice? The development of the sanitation institution in a world society*. Tampere, Finland: Tampere University of Technology.

SEI. (2005). *Linking water scarcity to population movements: From global models to local experiences*. Stockholm Environmental Institute (SEI): Stockholm.

Siebe, C., Chapela-Lara, M., Cayetano-Salazar M., Prado B., & Siemens, J. (2016). Effects of more than 100 years of irrigation with Mexico city's wastewater in the Mezquital Valley (Mexico). In H. Hettiarachchi & Ardakanian (Eds.) *Safe use of wastewater in agriculture: Good practice examples* (pp. 121–138). Dresden: United Nations University Institute for Integrated Management of Material Fluxes and of Resources (UNU-FLORES).

Transparency International. (2008). *Global corruption report 2008: Corruption in the water sector*. Cambridge, UK: Cambridge University Press. www.transparency.org/whatwedo/publication/global_corruption_report_2008_corruption_in_the_water_sector.

UN. (2015). *Transforming our world: The 2030 agenda for sustainable development*. New York: United Nations.

UNEP. (2015a). *Good practices for regulating wastewater treatment: Legislation, policies and standards*. Nairobi: United Nations Environment Program (UNEP). www.unep.org/gpa/documents/publications/GoodPracticesforRegulatingWastewater.pdf.

UNEP. (2015b). Options for decoupling economic growth from water use and water pollution. *Report of the international resource panel working group on sustainable water management*. Nairobi: United Nations Environment Program (UNEP).

UN-WATER. (2016). *Towards a worldwide assessment of freshwater quality: A UN-water analytical brief*. Geneva: UN-Water.

Voß, A., Alcamo, J., Bärlund, I., Voß, F., Kynast, E., Williams, R., et al. (2012). Continental scale modelling of in-stream river water quality: A report on methodology, test runs, and scenario application. *Hydrological Processes, 26,* 2370–2384.

WHO. (2006). *Guidelines of the safe use of wastewater, excreta and grey water—Vol. 2: Wastewater use in agriculture*. Geneva, Switzerland: World Health Organization (WHO). www.who.int/water_sanitation_health/wastewater/wwuvol2intro.pdf.

WHO. (2016). *Preventing disease through healthy environments: A global assessment of the burden of disease from environmental risks*. Geneva, Switzerland: WHO Press, World Health Organization. http://apps.who.int/iris/bitstream/10665/204585/1/9789241565196_eng.pdf.

WWAP. (2015). *The united nations world water development report 2015: Water for a sustainable world united nations world water assessment programme (WWAP)*. Paris, France: UNESCO.

WWAP. (2016). *The united nations world water development report 2016: Water and jobs. United nations world water assessment programme (WWAP)*. UNESCO: Paris, France.

The Opportunity Versus Risks in Wastewater Irrigation

Md Zillur Rahman, Frank Riesbeck and Simon Dupree

Abstract The impacts of climate change and human induced activities due to the development of urbanization, industries and agriculture are the biggest challenges in the field of water resource management in globally. In arid and semi-arid regions this issue of water scarcity is a great economic, environmental and social problem due to high water demand for food production. Thus, the demand of wastewater reuse has significantly increased to tackle the challenges due to water scarcity. As a result, on one hand, wastewater reuse has an enormous potential for agricultural use and economic development, on the other hand, there are significant environmental and health concerns. The objective of this chapter is therefore, to discuss the various perspectives and approaches of wastewater use in agriculture including the "fit-to-purpose" approach, which entails the production of treated wastewater that meets the needs of the intended end-users. Discussion on wastewater reuse in this chapter also focuses specially for the human safety of irrigation water containing microorganisms and microbial risks. Even after biological treatment, municipal wastewater still contains a large number of microorganisms (bacteria, viruses, parasites, worm eggs), including pathogens. Therefore, although there are plenty of opportunities for wastewater irrigation, yet a central aspect of water reuse is the possibility of transmission of infectious diseases.

Keywords Water scarcity · Climate change · Agriculture · Wastewater irrigation Water quality · Pathogens · Hygienic indicators

1 Water Scarcity: The Global Picture

The global impacts of climate change, increased urbanization and industrial development have amplified pollution and over-extraction of the world's freshwater resources. Similarly, the discharge of polluted water into the water bodies puts the

M. Z. Rahman · F. Riesbeck (✉) · S. Dupree
Humboldt University of Berlin, Berlin, Germany
e-mail: frank.riesbeck.1@agrar.hu-berlin.de

H. Hettiarachchi and R. Ardakanian (eds.), *Safe Use of Wastewater in Agriculture*,
https://doi.org/10.1007/978-3-319-74268-7_2

hydroecological systems under significant stress (Hamilton et al. 2006; Gosling and Arnell 2016; Zhou et al. 2017). Climate change in the past years has caused extreme conditions on a global scale such as strong storms with heavy precipitations on one hand, and long and dry periods of elevated temperatures on the other hand have impacted agricultural productivities and livelihoods (Gawith et al. 2017). One major consequence of these global changes is the constant reduction of fresh water availability worldwide. Water scarcity already affects every continent around the world. Around 1.2 billion people, or almost one-fifth of the world's population, live in areas of physical scarcity, and 500 million people are approaching this situation. Another 1.6 billion people, or almost one quarter of the world's population, face economic water shortage (Parekh 2016; Gray et al. 2016). However, water scarcity is both a natural and a human-made phenomenon. There is enough fresh water on the planet for people but it is distributed unevenly and too much of it is wasted, polluted and unsustainably managed. On the other hand, the following Fig. 1 shows the saline ocean water accounts 97% of all water on earth (Liu et al. 2011; Du Plessis 2017). According to Liu et al. (2011), the usable among all water on earth is just only 1% in which most is from groundwater (See Fig. 1). This indicates the limitation of the water resources that supply usable water for humans' use.

Table 1 summarizes the percent consumption of water by the domestic, industrial, and agricultural sectors in the different parts of the world especially in low- and middle-income countries and high-income countries (Du Plessis 2017). It is clear that the majority is being consumed in the agricultural sector (about 69%). For this reason, major acts to save water should be performed in this sector. A more efficient irrigation system can have a positive effect on global water availability. By the middle of the 21st century, as the world's population grows to around 9 billion, global demand for food, feed and fiber will nearly double, while increasingly, crops and landscape may also be used for bioenergy and other industrial purposes (Dale et al. 2016; Helander 2017). New and traditional demand for agricultural produce will thus put growing pressure on already scarce agricultural resources. While agriculture will be forced to compete for land and water with sprawling urban

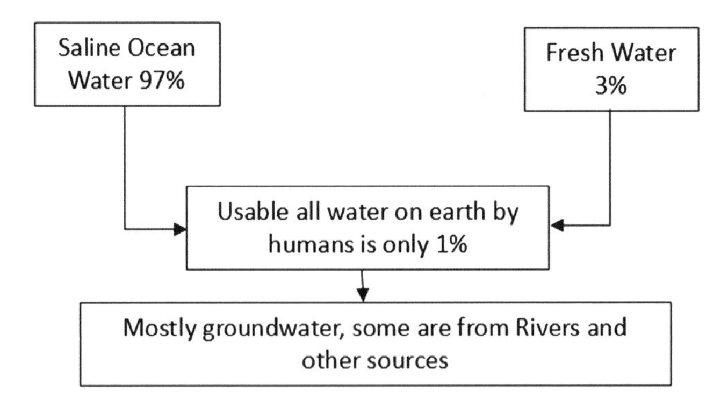

Fig. 1 Shows the available water for human use in worldwide that is mostly from groundwater

Table 1 (Constructed): Global water consumption by sector

69% of worldwide usage of water is for agriculture, mainly in the form of irrigation	22% for industrial purposes	8% for domestic purposes	1% for recreational use

Key water consumption locations:
- Agricultural water used for irrigation: Relatively higher in low- and middle-income countries, however, between 15 and 35% of the water that is withdrawn for irrigation purposes is used in an unsustainable manner
  - Asia is the highest of agricultural water usage because approximately 70% of the world's irrigated land occurs in Asia,
- Industrial water withdrawals: 5% in low-income countries and above 40% in some high-income countries

settlements, it will also be required to serve on other major fronts: adapting and contributing to the mitigation of climate changes, helping to preserve natural habitats, protecting endangered species and maintaining a high level of biodiversity.

The impacts of climate change and human induced activities due to the urbanization, industries and agriculture are the major challenges faced by the water resources management. Rising population, changes in dietary preferences and improved nutrition, increasing water withdrawals for urban, industrial, and recreational uses, and environmental protection have also greatly increased the global demand for good quality freshwater and its competition among users (De Fraiture et al. 2010; Harper and Snowden 2017).

With the world population approaching 9 billion people by 2050 the demand for food and clean water will be a critical concern in all countries, but will be particularly acute in areas that utilize irrigation for production of food, feed, and fiber. Global climate change may further exacerbate these problems through changing temperatures and long-term variations in annual precipitation amounts and regional rainfall distribution patterns. This situation is compounded by economic concerns including escalating energy costs, rising cost of inputs, persistent soil erosion and salination, increasing climate variability and continuing loss of environmental integrity (FAO 2017; Dale et al. 2016).

2 Agricultural Sustainability and Wastewater

Irrigated agriculture currently supplies 40% of the world food supply on approximately 20% of the arable land in agricultural production (Postel 1999). Irrigated crop production also provides much of the feed to sustain animals used for work or food in many parts of the world (Hagihara and Hagihara 2016). Because the world's arable land base is declining due to soil salinization, erosion and urbanization there is a need to substantially increase crop production in humid areas, while irrigated agriculture will be required to provide an even greater proportion of the food supply in both arid and humid regions (Riesbeck 2015, 2016c, d, 2017b). Freshwater diversions to irrigated agriculture are declining while the need to increase production is escalating rapidly. Subsurface drainage offers many benefits that will be necessary to meet these challenges in both humid and arid areas. In arid irrigated, as well as humid farming areas, removing excess water from the soil surface and the soil profile by natural and artificial means is a critical component in sustaining crop production. In many arid locations, irrigated crop production can only be physically sustainable if both soil salt concentrations and shallow groundwater elevations are adequately managed, whereas humid areas usually need field drainage to lower groundwater tables or intercept subsurface flows to prevent waterlogged conditions. Fortunately, compared to early irrigated civilizations, modern irrigation and drainage technologies in conjunction with advanced irrigation management options can control soil salinity and maintain sustainable crop production (Smedema et al. 2004; UNESCO 2009).

Due to climate change, the world is now experiencing longer drought periods and stronger rain and storm events. These cause gradual reduction in natural vegetation that used to help stabilize soil during water runoff, but with the absence of vegetation and stronger water runoff, soil is subjected to erosion by water and wind. Land degradation is a combination of several processes such as soil erosion, soil salinity, chemical contamination, desertification nutrient depletion and water scarcity (Table 2). Afforestation, toxic chemical soil contamination and soil salinity are an example of man-made causes for soil degradation that reduces available cropland for food production. So far 18% of the degraded land is cropland, 25% is central forests and 17% are north forests.

2.1 *Wastewater as a Solution*

The increasing water scarcity and water pollution control efforts in many countries have made treated municipal and industrial wastewater a suitable economic means of augmenting the existing water supply, especially when compared to expensive

Table 2 Key issues concerning irrigation water quality effects on soil, plants and water resources

Soil	Root zone salinity Soil structural stability Build-up of contaminants in soil Release of contaminants from soil to crops & pastures
Plants	Yield Salt tolerance Specific ion tolerance Foliar injury Uptake of toxicants in produce for human consumption Contamination by pathogens
Water resources	Deep drainage & leaching below root zone Movement of salts, nutrients & contaminants to groundwater & surface waters

Source Riesbeck (2016a, b, 2017a)

alternatives such as desalination or the development of new water sources involving dams and reservoirs.

Treated wastewater can be used for various non-potable purposes. The dominant applications for the use of treated wastewater (also referred to as reclaimed water or recycled water) include agricultural irrigation, landscape irrigation, industrial reuse, and groundwater recharge. Agricultural irrigation was, is, and will likely remain the largest reuse water consumer with recognized benefits and contribution to food security.

The majority of the water reused worldwide, is for agriculture, which is by far the largest water consumer. Due to increasing competition for water, farmers often have few alternatives to use raw or diluted wastewater to irrigate a range of crops (Qadir et al. 2013; Tessaro et al. 2016). In global context, wastewater use in agriculture is a growing phenomenon, especially where population densities are increasing and where freshwater is scarce due to increase consumption of goods and services (Mekonnen and Hoekstra 2016; Pfister et al. 2017). Thus, the reuse of wastewater is being widely supported as it helps address the global freshwater shortage issues. For instance, many developing countries (e.g. Argentina, China, Cyprus, Jordan, Mexico, Spain, Tunisia, and Saudi Arabia) as well as some water scarce regions in the developed world (such as Australia) are using wastewater for irrigation as a common practice (Devi 2009). In addition, wastewater can also be used for a variety of other purposes such as for industrial needs, urban and landscape irrigation, groundwater recharge, and wetland creation (Hamilton et al. 2006). There are three categories of wastewater that can be distinguished (UNW-DPC 2012):

- Direct use of treated wastewater
- Indirect use of untreated wastewater when water is abstracted from a river that receives wastewater. In this case the farmer might not be aware of the contamination
- Direct use of untreated wastewater, where fields are irrigated directly from sewage outlets.

Treated municipal wastewater can be used for all kinds of irrigation purposes as long as certain quality standards are met. Besides all benefits of wastewater use, it can also have adverse impacts on health and environment depending on the treatment level, type of irrigation and local conditions. It is important to mention that in several developing countries raw sewage is still being used for agricultural irrigation despite the established adverse effects on human health (Agyei and Ensink 2016). The total land irrigation with raw or partially diluted wastewater is estimated to be used at several million ha in fifty countries, which is approximately 10% of total irrigated land.

2.2 *Limitations and Risks*

Every anthropogenic use of water changes the amount and quality of available water resource in most cases with a negative impact caused by pollutants for further human use and aquatic ecosystems. Protecting and restoring the ecosystems that naturally capture, filter, store and release water, such as rivers, wetlands, forest and soils for agricultural food production, is crucial for increasing the availability of water of good quality.

Poorly controlled wastewater also means daily exposure to an unpleasant environment. The buildup of faecal contamination in rivers and other waters is not just a human risk: other species and the ecological balance of the environment are affected and threatened as well. The discharge of untreated wastewater into the environment affects human health by several routes:

- By polluting drinking water
- Entry into the food chain, for example via fruits, vegetables or fish and shellfish
- Bathing, recreational and other contact with contaminated waters
- By providing breeding sites for flies and insects that spread diseases.

Even after biological treatment, municipal wastewater still contains a large number of microorganisms (bacteria, viruses, parasites, worm eggs), including pathogens. Therefore, a central aspect of water reuse is the possibility of transmission of infectious diseases.

There have been implications of the transmission of many infectious diseases including cholera, typhoid, infectious hepatitis, polio, cryptosporidiosis, and ascariasis. Infectious agents are not the only concerns associated with wastewater; heavy metals, toxic organic and inorganic substances can also pose serious threats to human health and the environment—particularly when industrial wastes are added to the waste stream. For example, in some parts of China, years of irrigation with wastewater heavily contaminated with industrial waste is reported to have produced health issues such as enlargement of the liver, cancers, and raised rates of congenital malformation rates compared to areas where wastewater was not used for irrigation.

Despite the fact of the risk relationships of wastewater reuse for irrigation and agricultural activities, foodborne pathogen contamination of agricultural fresh products can also occur aside from irrigation wastewater (De Keuckelaere et al. 2015). For instance, post-harvest practice such as washing of fresh-cut produce, during further processing, or during preparation may also function as a means of cross-contamination (Holvoet et al. 2014; MacDonald et al. 2011). According to Harder et al. (2014), during processing time there are multiple sources for microbial contamination that include inadequate worker hygiene and poor handling practices, contaminated equipment, and wildlife. Vergine et al. (2015) emphasized that if fields are irrigated with treated municipal wastewater, the risk for human health occurs in the short-term period. The presence of pathogenic organisms in the topsoil and on the vegetables progressively reduced by natural processes (die-off, solar disinfection, rain, and transfer to the lower layers) and is influenced by various soil and environmental variables, such as soil texture, organic matter, pH, temperature, moisture content and nutrients. Thus, it is not only the issue of the risks of wastewater reuse in irrigation and pathogen contamination of agricultural fresh products, but the other factors are also responsible.

An important aspect of risk assessment in the wastewater reuse is the possible route of infection for humans (Riesbeck and Rahman 2015; Riesbeck 2016c, d, 2017c), that has shown in Fig. 2. Microorganisms are ubiquitous—that is, everywhere (worldwide) spread—due to wind, water and many others kinds of transport. It lives to about 100 trillion microorganisms in the human body. They contribute to essential physiological functions such as the structure of the immune system and the digestive system.

Only a small proportion of the microorganisms are pathogenic, i.e. these organisms can cause diseases in plants, animals or human (Table 3). Fecal-oral transmittable pathogens that are released with human and animal excreta, multiply not with a few exceptions in the treated wastewater. They can survive under favorable environmental conditions for days, weeks and some up to several months in the environment. For example, Bacteria 'Thermotolerant Coliforms' can survive in Freshwater and Sewage (<60 and usually <30 days), in crops (<30 and usually <15 days), in soil (<70 and usually <20 days). Protozoan Cysts 'Cryptosporidium oocysts' can survive in Freshwater and Sewage (<180 and usually <70 days), in crops (<3 and usually <2 days), in soil (<150 and usually <75 days). Worm eggs

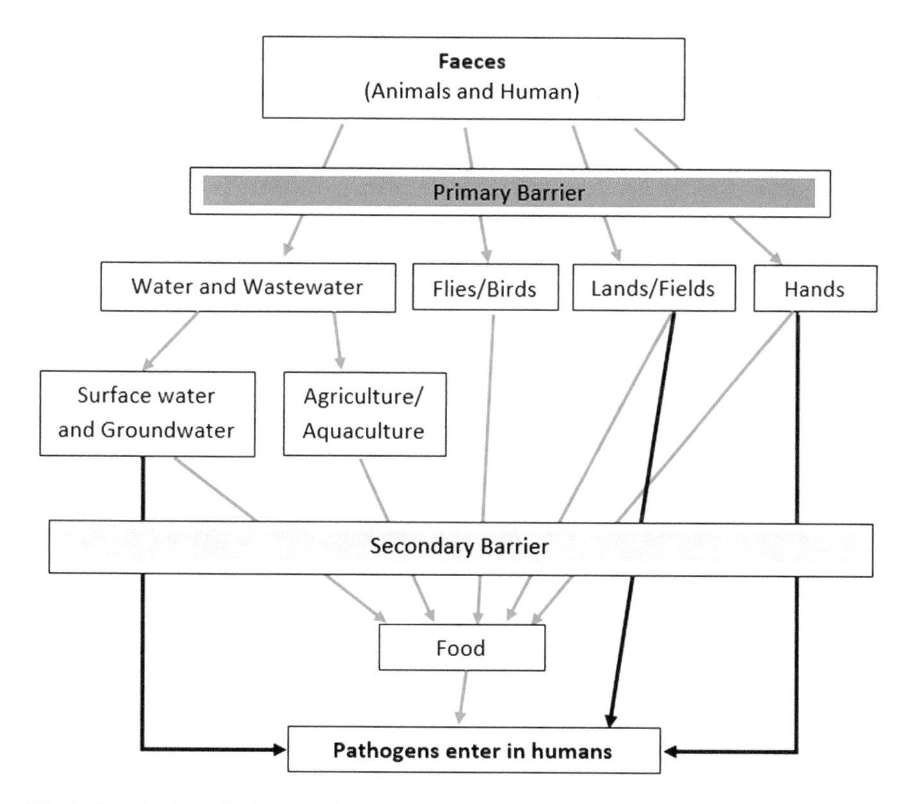

Fig. 2 Introduction of primary and secondary barriers to intervene the possible routes of infection for humans

Table 3 Some human and animal pathogens that can directly or indirectly cause disease via wastewater

Type	Pathogen	Diseases
Bacteria	*Salmonella* typhi	Typhus
	Shigella sp.	Bacterial dysentery
	Enteropathogene *Escherichia coli*	Enteritiden, enterotoxämien
	Pseudomonas aeruginosa	Dermatitis, Otitis
	Vibrio cholerae	Cholera
Virus	Polioviren	Meningitides
	Hepatitis A	Hepatitis Epidemic
Protozoe	*Entamoeba histolytica*	Amoebic dysentery
	Giardia lamblia	Lamblia dysentery
	Cryptosporidium sp.	Kryptosporidiose
Worms	*Ascaris lumbricoides*	Reel worm infestation
	Taenia sp.	Tapeworm infestation

Source Riesbeck and Rahman (2015); Riesbeck (2016d)

'Ascaris eggs' can survive years in Freshwater and Sewage and in soil, however it can survive in crops (<60 and usually <30 days).

Similarly, Nitrates from wastewater can build up to high concentrations in water sources underground. Nutrients may also cause eutrophication—undesirable excess in nutrients—in water sources. This can result in overgrowth of algae and harmful cyan bacteria. The toxins produced by some toxic cyan bacteria cause a range of effects on ecosystems and human health.

3 Hygienic Indicators

The variety of human pathogens and most of fecal origin of pathogens led to the development of the indicator concept (Mayer et al. 2016). As a rule, so-called indicator organisms used for microbiological water analysis. These serve as proof of faecal contamination in the water, since they always occur in the intestines of warm-blooded animals in large numbers and suggest that pathogens may be present.

They are as follows:

- Total Coliform bacteria, which only gives an indication of faecal contamination, because they come not only from the intestines of warm-blooded animals
- Faecal coliforms mainly *Escherichia coli* (*E. coli*), which may be regarded as evidence of fecal contamination, because they occur only in warm-blooded animals' intestines
- Fecal enterococci, which are also a proof of faecal contamination (more resistant than *E. coli*)

- Somatic coliphages can be used as indicator organisms for intestinal virus
- Clostridium perfringens as an indicator of pathogenic protozoa
- For worm eggs there are no indicators that they must be determined directly.

In general, irrigation water can be regarded as hygienically safe if it contains pathogens or substances or only in concentrations that humans and animals cannot damage.

The hygienic safety of surface water and treated waste water shall be demonstrated differently depending on the application (Mikola and Egli 2016; Kistemann et al. 2016; Schuster-Wallace and Dickson 2017). Studies required for classification of the water should be made for irrigation before and during its use. It is necessary to take them into account at the planning stage. Groundwater is usually hygienically safe. If there is reasonable suspicion of contamination, further investigations should be conducted. If there is suspicion of wastewater contamination or microbiological values listed are exceeded, additional tests may be required (see limit details in Table 4). Particular caution is required when the nature of surface water has the following criteria:

- Saprobic index (secure laboratory method) if not available than to consider following three options because of correlation,
- Ammonium content over 1 mg/l (NH_4)
- BOD5 about 10 mg/l (O_2)
- COD more than 60 mg/l (O_2)

4 Fit-to-Purpose Approach

An important new concept in water reuse is the "fit-to-purpose" approach, which entails the production of reclaimed water quality that meets the needs of the intended end-users. In the situation of reclaimed water for irrigation, the reclaimed water quality can induce an adaptation of the type of plant grown. Thus, the intended water reuse applications are to govern the degree of wastewater treatment required and, inversely, the reliability of wastewater reclamation processes and operation.

The main water quality factors that determine the suitability of treated wastewater for irrigation are pathogen content, salinity, sodicity, specific ion toxicity, other chemical elements, and nutrients. Local health authorities are responsible for establishing water quality threshold values depending on authorized uses and they are also responsible for defining practices to ensure health and environmental protection taking into account local specificities. The ISO guideline refers to factors involved in water reuse projects for irrigation regardless of size, location, and complexity. It is applicable to intended uses of treated wastewater in a given project, even if such uses will change during the project's lifetime; as a result of changes in the project itself or in the applicable legislation. The key factors in

Table 4 Hygienic-microbiological classification and application of irrigation water

Suitability class	Application	Number of colonies of fexal streptococci/ 100 ml (according to the directive for drinking water and bathing)[a]	Number of colonies of *Escherichia coli*[a]	*Salmonellae* 1000 ml (according to DN 38414-13)	Potentially infectious stages of human and pet parasites[b] in 1000 ml
1 (Drinking water)	– All greenhouse and field crops without restriction	Below limit of detection	Below limit of detection	Below limit of detection	Below limit of detection
2[c]	– Outdoor and greenhouse crops for raw consumption – School sports fields, public parks	$\leq 100^{d}$	$\leq 200^{d}$	Below limit of detection	Below limit of detection
3[c]	– Greenhouse crops not for human consumption – Field crops for raw consumption to fruit set; vegetables until 2 weeks before harvest – Fruits and vegetables for preservation (canned fruits/vegetables) – Grassland or green fodder until 2 weeks before cutting or grazing – All other outdoor crops without restriction – Other sports fields	≤ 400	≤ 2000	Below limit of detection	Below limit of detection
4[c, 5]	– Sugar, starch potatoes, oilseeds and non-food crops for industrial processing and seed until 2 weeks before harvest – Cereal until milk-ripe stage (not to be eaten raw) – Fodder for preservation until 2 weeks before harvest	Wastewater, which has at least passed through a biological treatment stage			– For intestinal nematodes no standard recommendation possible – For stages of *Taenia*: Below limit of detection

Source (DIN-19650 1965)

[a]Microbiological analyzes according to the usual procedures for bathing water

[b]To the extent necessary for safeguarding the health of humans and animals, an investigation of the proposed irrigation water on intestinal nematodes (*Ascaris* and *Trichuris* species and hookworms) and/or tapeworm life stages can be arranged (especially *Taenia*) according to WHO recommendation

[c]If a wetting on edible parts of the crop is prevented by the irrigation method, a restriction by microbiological hygiene suitability classes is not necessary

[d]Value should be as low as possible and "as required with a reasonable effort by the state of technology, taking into account the circumstances of the case" (according to directive for drinking water and bathing § 2 para. 3) During irrigation it must be ensured that staff and public are not harmed

[5]In case of spray irrigation, it has to be ensured through protective measures that employees and the public are not at risk

assuring the health, environmental and safety of water reuse projects in irrigation are the following according to ISO (ISO-16075-1 2015; ISO-16075-2 2015):

- meticulous monitoring of treated wastewater quality to ensure the system functions as planned and designed;
- design and maintenance instructions of the irrigation systems to ensure their proper long-term operation;
- compatibility between the treated wastewater quality, the distribution method, and the intended soil and crops to ensure a viable use of the soil and undamaged crop growth;
- compatibility between the treated wastewater quality and its use to prevent or minimize possible contamination of groundwater and surface water sources.

5 Summary

The use of wastewater is being widely supported as it helps address the global issues created due to the shortage of freshwater. It is evident that developing countries as well as in water scarce regions of the developed countries are equally paying more attention to wastewater as a solution to the water scarcity issues. Moreover, wastewater can be reused for a variety of purposes such as agricultural use, heavy industry uses, urban and landscape irrigation, groundwater recharge, and wetland creation. In general, treated municipal wastewater can be used for all kinds of irrigation purposes as long as the quality standards are met. The use of wastewater for agricultural productivities is significant; especially where population density is increasing while freshwater is getting scarcer. The "fit-to-purpose" approach, which entails the production of treated wastewater that meets the needs of the intended end-users, is becoming more practical. In the situation of wastewater irrigation, the reclaimed water quality can induce an adaptation of the type of plant grown. Thus, the intended water reuse applications are to govern the degree of wastewater treatment required and, inversely, the reliability of wastewater reclamation processes and operation.

References

Agyei, P. A., & Ensink, J. (2016). Wastewater use in urban agriculture: an exposure and risk assessment in Accra, Ghana. *Journal of Science and Technology (Ghana), 36,* 7–14.

Dale, V. H., Kline, K. L., Buford, M. A., Volk, T. A., Smith, C. T., & Stupak, I. (2016). Incorporating bioenergy into sustainable landscape designs. *Renewable and Sustainable Energy Reviews, 56,* 1158–1171.

de Fraiture, C., Molden, D., & Wichelns, D. (2010). Investing in water for food, ecosystems, and livelihoods: An overview of the comprehensive assessment of water management in agriculture. *Agricultural Water Management, 97,* 495–501.

de Keuckelaere, A., Jacxsens, L., Amoah, P., Medema, G., McClure, P., Jaykus, L.-A., et al. (2015). Zero risk does not exist: Lessons learned from microbial risk assessment related to use of water and safety of fresh produce. *Comprehensive Reviews in Food Science and Food Safety, 14,* 387–410.

Devi, M. G. (2009). *A framework for determining and establishing the factors that affect wastewater treatment and recycling.* Citeseer.

DIN-19650. (1999). *Hygienic concerns of irrigation water "Definition of suitability classes classes".* Deutsches Institut für Normung e.V.

Du Plessis, A. (2017). Global water availability, distribution and use. In *Freshwater challenges of South Africa and its Upper Vaal River.* Springer.

FAO. (2017). *The future of food and agriculture – Trends and challenges.* Rome.

Gawith, D., Hill, D., & Kingston, D. (2017). Determinants of vulnerability to the hydrological effects of climate change in rural communities: Evidence from Nepal. *Climate and Development, 9,* 50–65.

Gosling, S. N., & Arnell, N. W. (2016). A global assessment of the impact of climate change on water scarcity. *Climatic Change, 134,* 371–385.

Gray, J., Holley, C., & Rayfuse, R. (2016). *Trans-jurisdictional water law and governance.* Routledge.

Hagihara, Y., & Hagihara, K. (2016). Water resources conflict management: Social risk management. In *Coping with regional vulnerability.* Springer.

Hamilton, A. J., Versace, V. L., Stagnitti, F., Li, P., Yin, W., Maher, P., et al. (2006). Balancing environmental impacts and benefits of wastewater reuse. *WSEAS Transactions on Environment and Development, 2,* 117–129.

Harder, R., Heimersson, S., Svanström, M., & Peters, G. M. (2014). Including pathogen risk in life cycle assessment of wastewater management. 1. Estimating the burden of disease associated with pathogens. *Environmental Science and Technology, 48,* 9438–9445.

Harper, C., & Snowden, M. (2017). *Environment and society: Human perspectives on environmental issues.* Taylor & Francis.

Helander, H. (2017). Geographic disparities in future global food security: Exploring the impacts of population development and climate change.

Holvoet, K., de Keuckelaere, A., Sampers, I., van Haute, S., Stals, A., & Uyttendaele, M. (2014). Quantitative study of cross-contamination with *Escherichia coli, E. coli* O157, MS2 phage and murine norovirus in a simulated fresh-cut lettuce wash process. *Food Control, 37,* 218–227.

ISO-16075-1. (2015). *International Standard (ISO 16075-1): Guidelines for treated wastewater use for irrigation projects—Part 1: The basis of a reuse project for irrigation* (1st ed.).

ISO-16075-2. (2015). *International Standard (ISO 16075-2), 2015-part 2: Guidelines for treated wastewater use for irrigation projects—Part 2: Development of the project* (1st ed.).

Kistemann, T., Schmidt, A., & Flemming, H.-C. (2016). Post-industrial river water quality—Fit for bathing again? *International Journal of Hygiene and Environmental Health, 219,* 629–642.

Liu, J., Dorjderem, A., Fu, J., Lei, X., & Macer, D. (2011). *Water ethics and water resource management* (Ethics and Climate Change in Asia and the Pacific (ECCAP) Project, Working Group 14 Report). UNESCO Bangkok.

MacDonald, E., Heier, B., Stalheim, T., Cudjoe, K., Skjerdal, T., Wester, A., Lindstedt, B., & Vold, L. (2011). Yersinia enterocolitica O: 9 infections associated with bagged salad mix in Norway, February to April 2011. *Euro Surveill, 16.*

Mayer, R., Bofill-Mas, S., Egle, L., Reischer, G., Schade, M., Fernandez-Cassi, X., et al. (2016). Occurrence of human-associated Bacteroidetes genetic source tracking markers in raw and treated wastewater of municipal and domestic origin and comparison to standard and alternative indicators of faecal pollution. *Water Research, 90,* 265–276.

Mekonnen, M. M., & Hoekstra, A. Y. (2016). Four billion people facing severe water scarcity. *Science advances, 2,* e1500323.

Mikola, A., & Egli, J. (2016). Keeping receiving waters safe: The removal of PFOS and other micro pollutants from wastewater. *Proceedings of the Water Environment Federation, 2016,* 3602–3612.

Parekh, A. (2016). Journey of sustainable development by private sector actors. In *Water security, climate change and sustainable development.* Springer.

Pfister, S., Boulay, A.-M., Berger, M., Hadjikakou, M., Motoshita, M., Hess, T., et al. (2017). Understanding the LCA and ISO water footprint: A response to Hoekstra (2016) "A critique on the water-scarcity weighted water footprint in LCA". *Ecological Indicators, 72,* 352–359.

Postel, S. (1999). *Pillar of sand: Can the irrigation miracle last?* WW Norton & Company.

Qadir, M., Drechsel, P., & Raschid-Sally, L. (2013). *Wastewater use in agriculture.*

Riesbeck, F. (2015). *Irrigation and drainage—Worldwide techniques, technologies and management* (First draft). Study Report. Part II. "Drainage". Berlin, Germany: Humboldt Universität Berlin.

Riesbeck, F. (2016a). *Guideline for the management and evaluation of application of irrigation for Khuzestan province according to the recommendation of the German Association for Water, Wastewater and Waste (DWA)* (Study Report). Berlin, Germany: Humboldt Universität Berlin.

Riesbeck, F. (2016b). *IRRIGAMA—The web-based Information and Advisory system for environmentally friendly and economically efficient Control of irrigation use in Agriculture and Horticulture in Germany.* Berlin: Humboldt Universität Berlin.

Riesbeck, F. (2016c). *River management with the emphasis on water quality, qualified summary of the study "Impacts of pollutants on water quality"; "Source of pollutants, toxicology, risk factor identification and assessment—Total maximum daily load (TMDL)" (Second draft).* Berlin: Humboldt Universität Berlin.

Riesbeck, F. (2016d). *Water reuse assessment and technical review on the effects of disposing drainage water into the environment.* Berlin: Humboldt Universität Berlin.

Riesbeck, F. (2017a). *Irrigation and drainage management in Khuzestan with emphasis on water use efficiency—Decision support system (DSS) for irrigation and drainage management.* Berlin: Humboldt Universität Berlin.

Riesbeck, F. (2017b). *Overview of technique & technologies of wastewater treatments* (Final draft). Berlin: Humboldt Universität Berlin.

Riesbeck, F. (2017c). *River management with the emphasis on water quality, qualified summary of the study "Impacts of pollutants on water quality"; "Source of pollutants, toxicology, risk factor identification and assessment—Total maximum daily load (TMDL)"* (Final draft). Berlin: Humboldt Universität Berlin.

Riesbeck, F., & Rahman, Z. (2015). *Water reuse - ein Risiko für den Verbraucher?* Korrespondenz Wasserwirtschaft. DWA.

Schuster-Wallace, C. J., & Dickson, S. E. (2017). Pathways to a water secure community. In *The human face of water security.* Springer.

Smedema, L., Vlotman, W., & Rycroft, D. (2004). *Modern land drainage: Planning, design and management of agricultural drainage systems.* London: Taylor and Francis Group.

Tessaro, D., Sampaio, S. C., & Castaldelli, A. P. A. (2016). Wastewater use in agriculture and potential effects on meso and macrofauna soil. *Ciência Rural, 46,* 976–983.

UNESCO. (2009). *Water in a changing world* (The United Nations Development Report 3). Paris: UNESCO Publishing, und London: Earthscan: UNESCO.

UNW-DPC. (2012). Mid-term-proceedings on- capacity development for the safe use of wastewater in agriculture. In Ardakanian, R., Sewilam, H., & Liebe, J. (Eds.), *UN-Water decade programme on capacity development (UNW-DPC).*

Vergine, P., Saliba, R., Salerno, C., Laera, G., Berardi, G., & Pollice, A. (2015). Fate of the fecal indicator *Escherichia coli* in irrigation with partially treated wastewater. *Water Research, 85,* 66–73.

Zhou, Y., Ma, J., Zhang, Y., Qin, B., Jeppesen, E., Shi, K., et al. (2017). Improving water quality in China: Environmental investment pays dividends. *Water Research, 118,* 152–159.

The Impact of Wastewater Irrigation on Soils and Crops

Md Zillur Rahman, Frank Riesbeck and Simon Dupree

Abstract Wastewater used for agricultural irrigation covers wastewater of different qualities, ranging from raw and diluted, to those generated by various urban, industrial and agricultural activities. In general, wastewater use for irrigation is hardly depended on wastewater quality and soil conditions. Thus, the decision to practice wastewater irrigation should be based on pertinent soil and geologic properties as well as cropping intentions. This chapter aims to explore the impact of wastewater quality on soils and crops. The chapter also outlines the basic scientific requirements needed for successful use wastewater in this context.

Keywords Wastewater irrigation · Water quality · Soil properties
Nutrients · Salinity · Sodium hazard · Soil pH · Alkalinity

1 Introduction

In general, soil is a complex mixture and ideal place for plants as it holds nutrients and water. Therefore, it is important that the quality of soil is well protected for sufficient crops and plants productions. However, the increase of urbanization, industrial development and growing economic activities are contributing daily a high quantity of wastewater production. In many situations, these wastewaters are discharged either with or without proper treatment and sometimes are used for agricultural activities. Which at the end even treated, are affecting soil's physical, chemical and biological conditions as well as directly or indirectly affecting crops production.

However, the impact of wastewater on soil depends on the quality of wastewater as well as the characteristics of soil such as texture (sand, silt and clay), structure and pH of soil. Moreover, soil hydraulic conductivities, water retention capacity, water table and water infiltration rate in soil are also sensitive factors. For example,

M. Z. Rahman · F. Riesbeck (✉) · S. Dupree
Humboldt University of Berlin, Berlin, Germany
e-mail: frank.riesbeck.1@agrar.hu-berlin.de

H. Hettiarachchi and R. Ardakanian (eds.), *Safe Use of Wastewater in Agriculture*,
https://doi.org/10.1007/978-3-319-74268-7_3

in clay soil the water mobility will be low and similarly if the amount of heavy metals increase in soil, then pH of soil decreases.

Therefore, in order to effective utilization of the treated wastewater (TWW) for agricultural activities without further damage of land or underlying groundwater, site selection must be based on appropriate crop pattern, soil, hydrological and climate conditions and water quality (Lijó et al. 2017; Santos et al. 2017). In addition, surface runoff, groundwater movement, capillary rise and drift from irrigation spray play important role (see Fig. 1).

In this context, the main goal of this chapter is to explain the impacts of quality of wastewater (chemical, hygienic and physical characteristics). The physical and chemical properties of wastewater (quality factors of wastewater) play important role to determine the impacts on soil and crop production. In arid and semi-arid regions, the concentration of chemical properties is higher than in humid regions due to high evaporation rate.

There are number of properties in wastewater, however according to WHO (2006), the main components of wastewater that can affect soil and crops much are as:

- nutrients (nitrogen, phosphorus, potassium),
- salts, metal, pathogens,
- toxic organic compounds,
- organic matter,
- suspended solids,
- acid and bases (pH). In the following sections, some of these components are discussed.

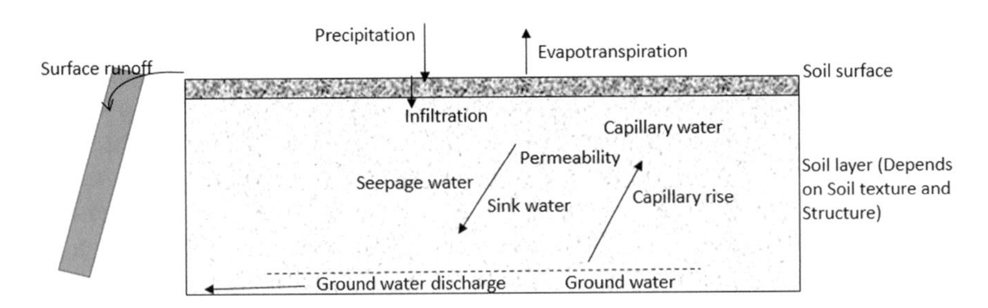

Fig. 1 Soil water properties (showing main factors only)

2 Nutrients Content

The TWW can contain nutrients, including other chemical elements, in higher concentrations than what we normally found in freshwater (ISO-16075-1 2015; ISO-16075-2 2015). However, it also contains macro elements such as nitrogen, phosphorus and potassium. In some cases, high nutrient loads can deteriorate the surface water quality when TWW is directly discharged into the environment (Auvinen et al. 2016). On the other hand, nutrients in TWW help farmers to reduce the use of chemical fertilizers, although, three major issues like quality, availability and timing need to consider before replacing conventional fertilizers with the chemicals found in the TWW (ISO-16075-1 2015, p10), for instance, one should ask:

- Quantity: Does the amount of nutrients provided by the TWW supply the needs of the plant?
- Availability: Can nutrients in the TWW be absorbed by the plants in the same way that nutrients normally supplied by the fertilizer are absorbed?
- Timing: Is the rate at which nutrients can be supplied during the season optimal for the crop?

Therefore, these three questions have highlighted the need of assessing the water-soil-crop relationship and the use of TWW under certain climate conditions. It is not only about the amount of nutrient's concentration in the TWW, but also related with cropping pattern and climate conditions of an area. In the following subsections, a brief overview of nitrogen, phosphorus and potassium is presented.

2.1 Nitrogen

When TWW is applied in an agricultural land, then Nitrogen in soil is increased because of additional Nitrogen from TWW. The organic nitrogen, ammonium (NH_4+) in TWW turn into nitrate (NO_3–N) by the nitrification process (ISO-16075-1 2015), which at the end can substitute farmers' need of nitrogen from TWW by replacing commercial fertilizers. Important to mention that, the process is also depended on cropping pattern, local climate conditions and soil conditions. Only a little part of nitrogen in soil in fact absorbed by the plants. In Table 1, the maximum level of total nitrogen in treated wastewater is shown. In order to improve soil's productivity, farmer must be careful that they cannot resistor the nitrogen concentrations when applying TWW in the agricultural land. Important aspect is the high nitrogen concentration that can reduce the salinity effects on crop production but can damage water quality if mixes with surface or groundwater water sources (ISO-16075-1 2015).

Table 1 Example of nutrients limit in treated wastewater for irrigation

Parameter	Units	Maximum value
Ammonium nitrogen	mg/l	30
Total nitrogen	mg/l	35
Total phosphorus	mg/l	7

Data Source (ISO-16075-1 2015)

2.2 *Phosphorus*

Most agricultural crops cannot remove full amount of phosphorus applied with secondary TWW. The additional phosphorus stores in the upper soil layers depending on soil properties such as soil's pH value. The pH value of soil also limits the mobility of phosphorus in soil and influence timing of application (ISO-16075-1 2015). The maximum value of total phosphorus in treated wastewater for irrigation is 7 mg/l (Table 1).

2.3 *Potassium*

Mobility of potassium in the soil is more limited than phosphorus. High concentration of potassium can also reduce the effects of salinity to crop production although the effects is less comparing to the effect of nitrogen (ISO-16075-1 2015).

3 Water Salinity and Sodium Hazard

Water salinity is one of the main factors for irrigation water quality issue and salinity refers the total amount of salts dissolved in the waterbody. Various ways agriculture water salinity can occur for example, saline water from rising groundwater and the intrusion of sea water into groundwater aquifers. In addition to salinity, TWW contains higher concentrations of inorganic dissolved substances such as total soluble salts, sodium, chloride, and boron. All of these can cause damage to the soil and the crop production (ISO-1 6075-1 2015).

Three main parameters are outlined in defining the quality of TWW with respect to salinity (ISO-16075-1 2015, p11). These are

- total content of salts due to osmotic effect,
- concentration of chlorides, boron, and sodium for its specific toxicity, and
- sodium adsorption ratio (SAR) due to soil permeability issues

Therefore, tolerance to salinity is different for different crops and yield reduction rates are also different according to salt tolerant thresholds of crops. Additional salts can affect plants by preventing an efficient water absorption due to higher osmotic pressure around the roots.

3.1 *Measurement of Salinity*

Salinity measurement is commonly expressed as Electrical conductivity (EC), which is a numerical expression for the ability of a medium to carry an electric current (Rhoades and Chanduvi 1999). In an aqueous solution, EC and total salt concentration are closely related, thus EC of water is used as a parameter to describe the total dissolved salt (TDS) concentration in water. EC is affected by temperature and it increases at a rate of approximately 1.9% per degree centigrade increase in temperature. Therefore, a reference temperature of EC is determined; 25 °C is most commonly used in this regard (Rhoades and Chanduvi 1999). The commonly used units for measuring electrical conductivity of water are: µS/cm (micro Siemens/cm) or dS/m (deci Siemens/m), where: 1000 µs/cm = 1 dS/m.

It is obvious that some plants and crops are more vulnerable to the electrical conductivity than each species has threshold of an electrical conductivity, beyond which yield is decreased. Examples of this relationship is presented in the Table 2.

Crops' salt tolerance capacity can be described by plotting its relative yield as a continuous function of salinity (Tanji and Kielen 2002), for example, Eq. 1. In case the irrigation water salinity exceeds the threshold for the crop, yield reduction occurs. The following equation provides an estimate of the yield potential as a function of the irrigation water salinity.

$$\%\text{Yield (of maximum)} = 100 - \text{b(ECe} - \text{a)} \quad (1)$$

where a = the salinity threshold expressed in dS/m; b = the slope expressed in percent per dS/m (loss in relative yield per unit increase in salinity) and ECe = the mean electrical conductivity of a saturated soil paste taken from the rootzone.

3.2 *Salinity Management*

Management practices for the safe use of TWW for irrigation primarily consists of the following steps (ISO-16075-1 2015, p19):

- selection of crops or crop varieties that will produce satisfactory yields under existing or predicted conditions of salinity or sodicity;
- special planting procedures that minimize or compensate for salt accumulation in the vicinity of the seed;
- irrigation to maintain a relatively high level of soil moisture and to achieve periodic leaching of the soil;
- use of land preparation to increase the uniformity of water distribution and infiltration, leaching and removal of salinity.

When the amount of salts increase in soil due to irrigation with TWW, then it is important to prevent the excessive accumulation of salt in the root zone. This can be

Table 2 Influence of the salt content (measured in terms of EC) on crops

Description	EC (mS/cm)	Salt content (%)	Influence on crops
Salt free	0–4	<0.15	Effect of salt negligible, just very susceptible crops affected
Slightly salty	4–8	0.15–0.35	Output losses in many crops
Moderate salty	8–15	0.35–0.65	Only tolerant crops are not influenced
Very salty	>15	>0.65	Only very tolerant crops are not influenced

Source (Riesbeck 2016)

done by applying more water for leaching the salt below the root zone; deeper into the soil, or out of crop growing field. This application of additional water is referred to as the "leaching requirement (LR), the fraction of infiltrated water that shall move through the root zone to keep salinity within acceptable levels" (ISO-16075-1 2015, p 19) (Fig. 2).

It is important to know how much to leach and when. Water demand for leaching salts from the soil essentially dependent on:

- The amount of salt present in the soil and groundwater
- Type of salts
- Quality of leaching water
- Water permeability of the soil
- Effectiveness of the irrigation system
- Target depth of desalination
- Applied leaching method

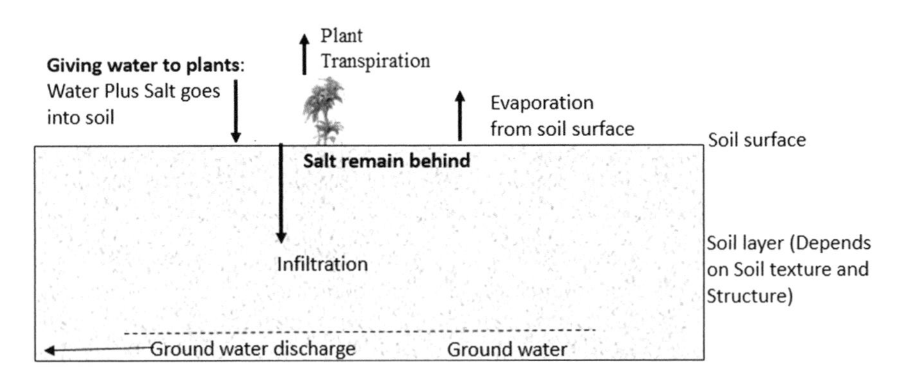

Fig. 2 Example of leaching process

Table 3 Salt tolerance of certain vegetables (*Source* Riesbeck 2016)

Total salt content [mg/l]		
Low (up to 500)	Moderate (up to 750)	High (up to 1000)
Beans, radish, garden radish, salad, carrot	Cucumber, onion, sweet pepper, tomato	Spinach, celery

Table 4 Tolerance to salinity of certain ornamental plants (*Source* Riesbeck 2016)

Total salt content [mg/l]			
Very low (150–250)	Low (up to 500)	Moderate (up to 750)	High (up to 1000)
Ferns	Ericaceae	Begonia	Chrysanthemum
Orchids	Gesneriaceae	Cyclamen	Carnation
Bromeliaceae	Araceae	Freesia	
Seed	Primrose	Rose	

Some examples of sensitivity and tolerance of agricultural plants to salinity are presented in Tables 3 and 4. As shown in Table 3, Spinach is one of the major vegetables to tolerate high level of salt contents (up to 1000 mg/l). Similarly, among ornamental plants like Chrysanthemum and Carnation have higher level of salt tolerance such as up to 1000 mg/l (Table 4).

3.3 Sodium Hazard

The damage caused to crops by sodicity of soils is even more common than the direct damage to crops as a result of salinity. This is because sodicity (adsorbed ion and electrolyte concentration) harms soil structure and water conductivity properties and thus creates improper water movement in the soil, leading to drainage and aeration problems (ISO-16075-1 2015). For example, fruit trees like avocado, citrus trees, and deciduous stone trees: plum, peach, apricot can be affected directly for extra concentrations of sodium in the TWW and due to sodium toxicity. Sodium Adsorption Ratio (SAR) parameter is used to determine the sodium hazard.

4 Soil pH Value and Alkalinity

Measuring the pH and alkalinity of irrigation water and soil solution is very importance as it can determine the success or failure of the crop cultivation (Dick et al. 2000; Stenchly et al. 2017). pH is a measure of acidic or basic (alkaline) in a solution by determining the hydrogen ion (H^+) activity in a given solution

(EPA 2006). The pH scale ranges from 0 to 14. Neutral water has a pH of 7.0; water measuring under 7.0 is acidic; and that above 7.0 is alkaline or basic. According to EPA (2006), several other factors also determine the pH of the water, including:

- bacterial activity;
- water turbulence;
- chemical constituents in runoff flowing into the waterbody;
- sewage overflows; and
- impacts from other human activities both in and outside the drainage basin.

Furthermore, nutrient deficiencies can occur when both pH is too high (above 7.0) and high Alkalinity. The alkalinity of water is related to pH and it measures water's ability to neutralize acidity (UMASS 2018). In other words, alkalinity is the buffering capacity of the water (Turner 2017), with range of pH between 7.0–14.0. Alkalinity is usually expressed as ppm or mg/L of Calcium Carbonate ($CaCO_3$). The main components causing the water alkalinity are:

- Carbonates (CO_3^{-2})
- Bicarbonates (HCO_3^{-})
- Soluble hydroxides (OH^{-})

Thus, farmers must be aware that low pH might result in micronutrient toxicities and damage the plant's root system. High alkalinity such as bicarbonates (HCO_3^{-}) and carbonates (CO_3^{-2}) can clog the nozzles of pesticide sprayers and drip tube irrigation systems (UMASS 2018). It is important for the farmers to monitor and control pH and alkalinity specially when using TWW for agricultural activities. Sometimes, many farmers have to add acid to their irrigation water and need to find how much acid to add by using the values of pH and alkalinity. Adding acid actually means adding hydrogen ions. So, it is obvious that both pH and alkalinity are essential for finding the correct amount of acid that have to add to the irrigation water in order to reach the required pH.

5 Toxic Ions, Heavy Metals, and the Suspended Solids in Wastewater

The irrigation water quality can also be determined by toxicity of specific ions (Elgallal et al. 2016; Alemu and Desta 2017), in which some of the ions can damage the agricultural crops those irrigated by TWW. The most common ions in wastewater which might cause a toxicity problem are Chloride, Sodium and Boron.

Chloride is an essential nutrient for plants and serves to transport cation in the plant for cell hydration. In the case of TWW irrigation, Chloride toxicity is more common and many plants are sensitive to it. Similarly, Boron is an essential element for plant growth and plant processes like cell division, elongation and nucleus acid metabolism. However, in the case of TWW irrigation is used, Boron is common (ISO-16075-1 2015). In excess amount, Boron becomes toxic for plants and toxicity occurs in very low concentrations. For example, Boron concentration level for Citrus and Blackberry is <0.5 mg B/l, contrary, Boron concentration level for Millet, Tomato, Alfalfa, Parsley, Beets, Sugar beet and Cotton is >4 mg B/l (Riesbeck 2016).

Therefore, farmers have to make sure that irrigation water from TWW is suitable from toxic ions (considered level of toxicity). This can be achieved by proper leaching, less use of fertilizers that contain Chloride or Boron or Sodium, selecting the right crops and good agricultural practice can help to avoid the damage of soil and crop production from toxicity.

Furthermore, wastewater with high heavy metal concentrations is not suitable for agricultural production (Qureshi et al. 2016; Makoni et al. 2016, Woldetsadik et al. 2017). Heavy metals (such as cadmium, zinc, lead, nickel, copper, platinum, silver, titanium and so on) are known for:

- influencing plant growth negatively,
- accumulating in plants and reach the human food chain, and
- accumulating in the soil.

Finally, in respect to suspended solids, although suspended solids in wastewater do not have much impacts on plants, however they play an important role for sustainable irrigation systems. For instance, suspended solids can damage some systems when using modern irrigation technologies such as sprinkler systems, drip irrigation, as well as pumping in pipe systems. High quality irrigation water is essential for drip irrigation systems and suspended solids can create major problems and damage the systems.

6 Concluding Remarks

Wastewater can be a promising alternative to irrigation water; however, it is important to understand the scientific context as well as the conditions. The chapter has discussed and outlined the basic scientific requirements needed for successful wastewater irrigation. Specially the relation of water quality, soil characteristics, and how the quality of the irrigation water affects both crop yields and soil physical conditions need to be taken into consideration. Moreover, different crops require different irrigation water qualities. Therefore, testing the irrigation water prior to selecting the site and the crops to be grown is critical. The quality of some water sources may change significantly with time or during certain periods.

References

Alemu, M. M., & Desta, F. Y. (2017). Irrigation water quality of River Kulfo and its implication in irrigated agriculture, South West Ethiopia. *International Journal of Water Resources and Environmental Engineering, 9,* 127–132.

Auvinen, H., Du Laing, G., Meers, E., & Rousseau, D. P. (2016). Constructed wetlands treating municipal and agricultural wastewater–An overview for flanders. *Natural and Constructed Wetlands.* Belgium: Springer.

Dick, W. A., Cheng, L., & Wang, P. (2000). Soil acid and alkaline phosphatase activity as pH adjustment indicators. *Soil Biology & Biochemistry, 32,* 1915–1919.

Elgallal, M., Fletcher, L., & Evans, B. (2016). Assessment of potential risks associated with chemicals in wastewater used for irrigation in arid and semiarid zones: A review. *Agricultural Water Management, 177,* 419–431.

EPA. 2006. *Voluntary Estuary Monitoring Manua: Chapter 11: pH and Alkalinity* (2nd Edn.).

ISO-16075-1. (2015). International Standard (ISO 16075-1): Guidelines for treated wastewater use for irrigation projects—Part 1: The basis of a reuse project for irrigation, First edition.

ISO-16075-2. (2015). International Standard (ISO 16075-2), 2015-part 2: Guidelines for treated wastewater use for irrigation projects—Part 2: Development of the project, First edition.

Lijó, L., Malamis, S., González-García, S., Fatone, F., Moreira, M. T., & Katsou, E. (2017). Technical and environmental evaluation of an integrated scheme for the co-treatment of wastewater and domestic organic waste in small communities. *Water Research, 109,* 173–185.

Makoni, F. S., Thekisoe, O. M., & Mbati, P. A. (2016). Urban wastewater for sustainable urban agriculture and water management in developing countries. *Sustainable Water Management in Urban Environments.* Springer.

Qureshi, A. S., Hussain, M. I., Ismail, S., & Khan, Q. M. (2016). Evaluating heavy metal accumulation and potential health risks in vegetables irrigated with treated wastewater. *Chemosphere, 163,* 54–61.

Rhoades, J. D., & Chanduvi, F. (1999). *Soil salinity assessment: Methods and interpretation of electrical conductivity measurements.* Food & Agriculture Organization.

Riesbeck, F. (2016). Guideline for the management and evaluation of application of irrigation for Khuzestan province according to the recommendation of the German association for water, wastewater and waste (DWA). Study Report. Berlin, Germany: Humboldt Universität Berlin.

Santos, S. R., Ribeiro, D. P., Matos, A. T., Kondo, M. K., & Araújo, E. D. (2017). Changes in soil chemical properties promoted by fertigation with treated sanitary wastewater. *Engenharia Agrícola, 37,* 343–352.

Stenchly, K., Dao, J., Lompo, D. J.-P., & Buerkert, A. (2017). Effects of waste water irrigation on soil properties and soil fauna of spinach fields in a West African urban vegetable production system. *Environmental Pollution, 222,* 58–63.

Tanji, K. K., & Kielen, N. C. (2002). *Agricultural drainage water management in arid and semi-arid areas.* FAO.

Turner, A. B. (2017). *Measuring inorganic carbon fluxes from carbonate mineral weathering from large river basins: The Ohio River Basin.*

UMASS. (2018). Water Quality: pH and Alkalinity [Online]. Available: https://ag.umass.edu/greenhouse-floriculture/fact-sheets/water-quality-ph-alkalinity (Site visited 24/01/2018).

WHO. (2006). *Guidelines for the safe use of wastewater, excreta and greywater.* World Health Organization.

Woldetsadik, D., Drechsel, P., Keraita, B., Itanna, F., & Gebrekidan, H. (2017). Heavy metal accumulation and health risk assessment in wastewater-irrigated urban vegetable farming sites of Addis Ababa, Ethiopia. *International Journal of Food Contamination, 4,* 9.

Selecting the Treatment Technology for Wastewater Use in Agriculture Based on a Matrix Developed by the German Association for Water, Wastewater, and Waste

Roland Knitschky and Hiroshan Hettiarachchi

Abstract Treatment of wastewater for the purpose of reuse is a complex task. In addition to the national and international regulations and standards on water quality and treatment technology, there are many other constraints that need to be taken into account, such as the financial resources and the level of training of local operating personnel. In order to methodically simplify the selection process, an assessment tool was developed by the German Association for Water, Wastewater and Waste (DWA) in 2008. This assessment tool is presented as a matrix (DWA Matrix, hereafter) that takes a variety of wastewater treatment processes into account. Within the DWA Matrix, each step in a process is assessed with regard to a diverse number of aspects, such as, discharge quality, costs, consumption of material and energy, expenses for preventative maintenance, and so forth. The assessment conducted on individual treatment methods allows them to be compared with each other and gives information about the risks of individual processes related to the water reuse. The objective of this chapter is to present background information on the process, and then to discuss how the DWA Matrix can be used for water reuse applications specifically in agriculture.

Keywords Agriculture · Assessment · Irrigation · Selection criteria Treatment techniques · Water reuse · Wastewater treatment

R. Knitschky (✉)
German Association for Water, Wastewater and Waste (DWA), Hennef, Germany
e-mail: KNITSCHKY@dwa.de

H. Hettiarachchi
United Nations University (UNU-FLORES), Dresden, Germany
e-mail: hettiarachchi@unu.edu

H. Hettiarachchi and R. Ardakanian (eds.), *Safe Use of Wastewater in Agriculture*,
https://doi.org/10.1007/978-3-319-74268-7_4

1 Introduction

Wastewater is a resource. Using this resource is already a common practice in some water-stressed countries. In the future, the use of treated wastewater will be an essential component for a sustained water resources management plan in many countries, and it may also become an important component of climate change adaptation. The reuse of water can address water shortage issues created by steadily rising water consumption and limited water resources very well. The needed level of treatment of wastewater, however, has to be decided based on the economics and the requirements all planned activities and potential risks related to the water. The tightening of environmental legislation in many countries (e.g., Australia, Jordan, and the USA) together with new reuse guidelines have given strong impetus to the proper reuse of water over the past 20 years (DWA 2008).

Recent research indicates an increasing trend of regulated reuse projects (Asano 2007; AQUAREC 2006; Jimenez and Asano 2008). However, there is still limited availability of data on the share of water reuse within the global water consumption. Figure 1 shows the reused volume for the largest known re-users of treated wastewater worldwide in the year 2008. The potential scope of water reuse applications is very wide: the largest water need arises from irrigation in agriculture, followed by industrial uses and other applications in urban/tourist sectors—with urban applications mostly referring to use in green areas and street cleaning (Asano 2007; Jimenez and Asano 2008). However, regulated water reuse continues to play only a marginal role in the total water demand.

Many different techniques are available nowadays for treatment of wastewater. However, information on how they should be selected and on what conditions is not readily available. The German Association for Water, Wastewater and Waste (DWA) took the lead on filling this gap by publishing a technical report on Treatment Steps for Water *Reuse* (DWA 2008). While the objective of the DWA

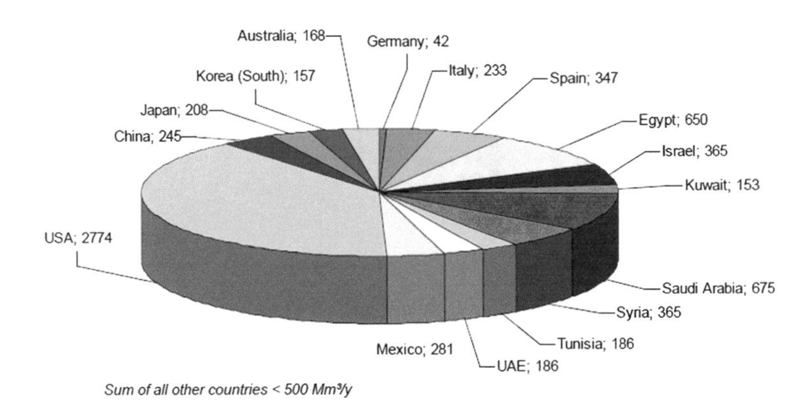

Fig. 1 Reused volume (10^6 m^3) for the largest users of treated wastewater worldwide (Kompetenzzentrum Wasser Berlin 2012, as cited in Fuhrmann et al. 2012)

report was not limited to reuse in agriculture, the objective of this present publication is to specifically address agricultural applications. The next sections aim to provide a guide on how to make a first selection of appropriate treatment technologies. In addition, they will explain how the DWA Matrix can help any organization in the field of wastewater management to get into the Safe Use of Wastewater in Agriculture (SUWA) business.

2 Water Reuse in Agriculture

Agriculture is by far the largest water consumer on a global scale, as illustrated in Table 1, and the demand is increasing. Therefore, use of wastewater to offset part of that need is logical, and it is already being practiced. Over 20 million hectares of agricultural land is currently irrigated with already used water (Hettiarachchi and Ardakanian 2016), and it continues to represent a large potential for growth.

In many developing countries and regions, use of non- or insufficiently treated wastewater in agriculture is very common. In particular in urban or peri-urban regions raw wastewater from the local population is being used for irrigation not only because it is available for free but also due to the nutrient content. In addition, the supply is relatively steady. The combination of these three factors also leads to high potential for water reuse in agricultural applications. Many developing countries and regions have introduced quality standards for the reuse of water (mainly on the basis of relevant international directives or guidelines, see Sect. 3.1). However, there is less regard on the actual implementation of these guidelines in practice.

Wastewater is the raw material needed to produce the product "adequately treated wastewater". This product should have specific qualities depending on the intended end use. For example, the permitted nutrient content depends on the vegetation, season, and the soil conditions. The hygienic aspects, on the other hand, may depend on the irrigated agricultural products and the method of cultivation. Similarly, the solid matter content should depend on the type of irrigation.

For sustainable water resources management it is essential to recognize wastewater as an important resource. However, the corresponding treatment and monitoring of the usage is indispensable in order to minimize all the risks associated with the water reuse. The objective of that treatment is to make reclaimed wastewater a secondary resource, which is fit for a specific use in agriculture.

Table 1 Competing water uses (Source of Data: United Nations 2003)

Type of Use	Global level (%)	High-income Countries (%)	Low-income Countries (%)
Agricultural	70	30	82
Industrial	22	59	10
Domestic	8	11	8

3 What Aspects to Consider in Selecting Treatment Techniques

Requirements for wastewater treatment for reuse go beyond the needs of a typical wastewater treatment facility, and the process may also require additional treatment steps. Other challenges can also emerge through the interplay of the continuous inflow of wastewater and often discontinuous consumption of the treated water. As a result, storage and plans for usage should also be taken into consideration (Fuhrmann et al. 2012). This may demand for storage capacity, which can be arranged both in surface storage tanks and also through deliberate storage in aquifers. Storage of water, however, may result in other quality demands—such as nutrient removal while using aquifer storage,—and in different quality issues such as microbial recontamination.

For the safe reuse of water, the following well-established treatment steps are typically taken into consideration (as displayed in the DWA Matrix, see Appendix B):

- Mechanical treatment, such as, sieving/screening and sedimentation,
- Biological treatment, such as, the activated sludge process, trickling filters, wastewater ponds, up-flow anaerobic sludge blanket (UASB) reactors, helophyte treatment plants or constructed wetlands,
- Combined wastewater storage- and treatment tanks,
- Filtration, sedimentation/flocculation, membrane technology, and
- Disinfection.

The technical processes of the above mentioned treatment steps are well established in general. What is less known is, how the local conditions may pose complex challenges to the implementation of water reuse infrastructure and its reliable operation. Success of a water reuse project depends on how well the treatment steps are selected and combined. There is also a need to be clear about the specific reuse-bound requirements. Beside the technological aspects, ecological, institutional, economic, and social aspects need to be taken into account. While social aspects are covered in detail in another chapter of this book, some of the mentioned key aspects to be considered in decision making are briefly discussed in the next few sections.

3.1 Health and Environmental Aspects

Municipal wastewater often contains substances that can trigger health-hazards even after conventional treatment. The most common examples are human pathogenic microorganisms in the form of bacteria, viruses, parasites, and helminth eggs and the remains of persistent chemical substances (AQUAREC 2006; WHO 2006; USEPA 2004). As a general rule, appropriate disinfection procedures should be employed to ensure the pathogens are reduced to acceptable limits through removal, destruction, or inactivation. Harmful inorganic salts and persistent anthropogenic organic substances have to be limited as well.

Similar to all other water uses, safe use of treated wastewater in agriculture also demands for a certain assurance of the water quality. However, the expected minimum requirements for the water quality may differ from application to application.

For safe agricultural irrigation applications specific water quality standards should be set by the responsible regulators. Information and recommendations from established international guidelines, such as, ISO 16075-1, 2 and 3 (ISO 2015), WHO Guideline for Water Reuse (2006), FAO Irrigation and Drainage Paper 29 on "Water quality for agriculture" (Ayers and Westcot 1985) should be taken into consideration as much as possible. Some of these international guidelines focus on risk-based and multi-barrier approaches, which requires a demanding implementation process (compared to a "simpler" definition of standards for water quality). However, at the same time, water-reuse standards should be harmonized within the national system of water and health regulations, and they should not hinder the potential of water reuse, for example, by using the containing nutrients for plant growth. National standards such as the German DIN 19650 "Irrigation—hygienic concerns on irrigation water" (DIN 1999), California Code of Regulations "Title 22" (CCR 2015), USEPA Guidelines for Water Reuse" (USEPA 2012) may serve as further references. European standards for water reuse are currently being developed.

Selected treatment technology needs to comply with the health and safety requirements to safeguard the (a) operating personnel at the treatment facility, (b) farmers who use treated wastewater, and (c) consumers of the products grown with treated wastewater. Further measures may be needed in particular with respect to health-risk awareness and epidemiological aspects. Other issues such as odor and aerosols may also have influence on health aspects.

Health aspects must be taken into consideration not only during the selection of the treatment technology and the operation of facilities but also within the complete process of water reuse in agriculture by even tracking down to the bottom of the production chain.

3.2 *Financial Sustainability*

Making treated wastewater available for agricultural applications is attractive; but it does come with a price tag. Water production will generate investment and operational costs. Appendix A.2 and A.3 gives a detailed tabulation. The resulting tariffs can be used as an argument in favor of such practices as long as they are lower than what is usually paid for comparable groundwater and surface water, including energy costs for delivery.

The management of water demands via appropriate prices for different types of use, such as, potable, domestic, industrial, and irrigation purposes, can contribute to a more effective use of water (Fuhrmann et al. 2012). In the same way one may encourage innovative solutions with closed water cycles for rural and urban areas. The principles of the European Water Framework Directive, therefore, demand a financial contribution from both the consumer and also from the polluter (DWA 2008). As per DWA (2008), socially acceptable and progressive tariffs set according to the ability and willingness of the user to pay are to be differentiated politically. They should also be regularly adjusted for inflation, in order to secure the necessary funding for the facility operation and customer services. In the long run, a high percentage of costs should be covered to ensure the economic sustainability of water reuse projects.

Ideally, the operation and maintenance costs should be covered by beneficiary users. Investors' own capital, state subsidies, and/or loans can be used for funding a new water reuse project. Development banks usually look at feasibility studies, which examine alternative concepts and technologies and illustrate inexpensive solutions both for the investors (low investment or operating costs) and the users (suitable tariffs) to provide funds.

There are numerous examples of well-coordinated and integrated water reuse projects that have illustrated ways for economically sustainable investments through the application of adopted frameworks, regulations, and standards—hence, ultimately through state regulation. Some examples can be found in AQUAREC (2006), EMWIS (2007), and Lazarova et al. (2013). Consumers in water-scarce countries and regions, such as Singapore, South Africa, Australia, and California, have adapted themselves, in the mid- and long-term, to the regionally available water resources with thoroughly varied water quality and different prices (DWA 2008).

3.3 Operational Aspects

Even the best technology bears considerable risks. These risks emerge when the process of water treatment, storage, distribution, and application cannot be executed as intended, for reasons beyond technological constraints. Apart from good equipment and technology, there is also a need for trained employees. Depending on the complexity of the agricultural reuse system, the water treatment processes used, as well as the operation and maintenance of the infrastructure, it also requires corresponding system management expertise.

Due to the sensitivity of the health aspects, personnel involved in the process should be able to act responsibly. Thus, the recruitment of suitable operating personnel is important. The personnel involved need to maintain the required qualifications through tailor-made training measures. Continuous follow-up trainings and examinations are recommended, especially in the early years after implementation of a water reuse project. However, in some countries and regions, these requirements are often contrary to the realities due to various reasons, such as (DWA 2008; Fuhrmann et al. 2012):

- Unclear institutional responsibilities,
- Strong hierarchical and centralized management structures with limited possibilities for decisions on-site,
- Inadequate budget for operation and maintenance,
- Lack of sufficient furnishing with operating resources, in particular equipment, spare parts, tools, energy, and chemicals,
- Personnel with insufficient qualification and limited possibilities for further training,
- Poor wages/salaries that do not motivate employees,
- Unmet demand for improvement of the image of employees (from "Sewer operator" to "Resource manager").

These conditions present enormous challenges to the project implementation and the success of investments in water reuse projects depends on how they are addressed. Appendix A.5 gives a brief overview of requirements on operating personnel.

3.4 Technological Aspects

The technologies selected for agricultural use of treated wastewater should be able to address the following (Fuhrmann et al. 2012): hygienic aspects (protection of health), biologically degradable substances (avoidance of odors), inorganic substances (protection against salinity), nutrients (protection against over-fertilization) and concentrations of solids (with regard to blockage of irrigation systems). For economic reasons, however, the selection of technology should target only the degree of treatment necessary to meet the minimum requirements applied to the expected irrigational application. Extensive reference examples on the selection of suitable water treatment and distribution technologies can be found in the literature (AQAREC 2006; Asano 2007; DWA 2008; Lazarova et al. 2013).

The treatment requirements for reuse purposes go beyond the main expectation of a typical wastewater treatment facility, which is to eliminate solids, organic matter and nutrients (see Fig. 2). The intended use of the water for irrigation may require additional treatment steps especially due to hygienic aspects (Annex A.1), the nutrients content (Annex A.6) and the concentration of solid matter (Annex A.7).

The technologies that are recommended for a controlled treatment for water reuse are already mentioned above (DWA 2008). All mentioned treatment processes are of relevance for the various purposes of reuse, and all are well established in general. Each technology has specific characteristics and functions in a treatment process; some can be seen alternating, some aid and abet further steps. At the end, in most cases it is a combination of treatment steps that achieves the desired result. But some details and characteristics may pose complex challenges to the implementation of

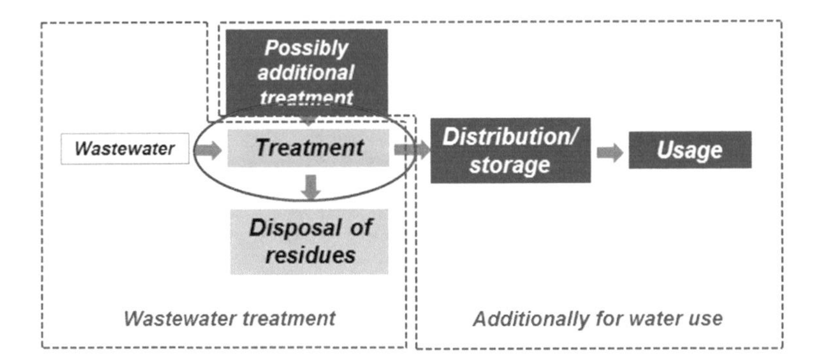

Fig. 2 System boundary of conventional wastewater treatment and additional aspects for water reuse (Firmenich et al. 2013)

water reuse infrastructure for irrigation purpose and its reliable operation. The degree of mechanization, robustness, process stability, the ability to influence the discharge quality operationally, and the accumulation of residues are only a few of these challenges (Annex A.6). The DWA Matrix in the Annex B gives important orientation and overview.

4 Selection of the Treatment Technique

Wastewater treatment with the aim of water reuse should be carried out using the technique best suited to the individual case of application. For the selection of the treatment technique, the variation of each constraint within the local conditions should be taken into consideration. In general, all aspects introduced in the previous sections should be given the due consideration. With these two sets of information in hand, the next question is what would be the best way to manage the decision-making process.

The selection process is of relevance for all stakeholders of a reuse project. It also involves financial, operational, quality and risk management aspects. Therefore, the decision-making process needs to be methodical, logical, and efficient. To organize the decision-making process, a tool was developed by the DWA Working Group on Water Reuse BIZ-11.4 (DWA 2008) in the format of an assessment matrix.

This matrix (DWA Matrix) gives a help to planers, designers, authorities, and even users in the primary decision-making phase of a project and allows a rational orientation in further improvement phases. Therefore, it essentially provides a general assessment of available options that can be used as a basis for further investigations to incorporate the local conditions. The DWA Matrix supports transparency in technologies and facilitate useful and reasonable decisions even in the case, that the expert's knowledge is limited. The Matrix explicitly will not replace engineers' assessments and tailor-made decisions.

The DWA Matrix has been developed to address water reuse needs in general, even though the present publication focuses only on agricultural irrigation. It is intended to cover a wide range of areas of application including urban uses (e.g., irrigation of parks, street cleaning, fire-protection) and non-potable domestic purposes (e.g., toilet flushing). Potable and industrial water use as well as alternative disposal concepts based on separation of sewage streams are excluded in this edition of the DWA Matrix. Indirect reuse and recharge into aquifers will be taken in account by the DWA Working Group on Water Reuse BIZ-11.4 in a further edition of the guideline expected within the next years.

The DWA Matrix presents various process steps of water treatment and provides the user an opportunity to compare/assess process steps with regard to various aspects, such as, discharge quality, costs, consumption of materials and energy, expenditure for preventative maintenance, and so on.

5 Structure of the DWA Matrix

Figure 3 below shows how the elements of the DWA assessment matrix are organized. Table 2 presents a snapshot of what is included in the first column of the DWA Matrix displayed in Appendix B. These are the criteria presented in Sect. 3 as the key aspects to be considered in decision-making. Each aspect is subdivided based on its nature and other requirements. This ultimately breaks the column 1 down to 44 lines (Table 2). All line items are clearly defined in Appendix A. The next columns of the DWA Matrix contain various technical options and process steps, one after another, of wastewater treatment. The complete assessment matrix shown in Appendix B is divided into the following five thematically grouped technologies: (a) mechanical treatment, (b) treatment ponds and tanks, (c) biological processes with higher requirements on operating personnel, (d) filtration and flocculation process steps, and (e) options for disinfection.

The assessment is facilitated in categories such as "high", "medium", and "low," and is partly supplemented by specific key data, such as, energy consumption or degree of elimination of specific wastewater parameters. The details are based on evaluations of the sources given in the references as well as the expert opinion of DWA Working Group BIZ-11.4 (DWA 2008). The number(s) presented immediately next to each field indicate the relevant source(s) and the details are presented in the legend provided at the end of Appendix B.

6 Summary

Awareness of the potential of water reuse is increasing internationally. The topic represents a complex but rewarding task, which, beyond the technical questions of wastewater treatment, has to take many other different aspects and implications into

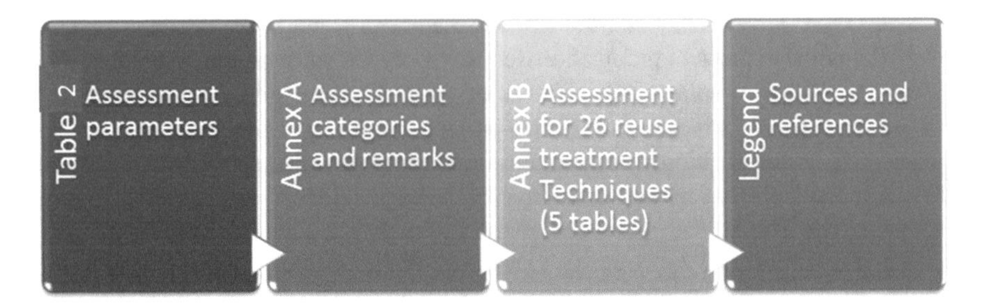

Fig. 3 Elements of the DWA assessment matrix "Treatment steps for water reuse"

account. Water scarcity has also created a growing market for water reuse, especially in agricultural irrigation. It is necessary to implement additional infrastructure and technology, not only for the treatment of wastewater, but also for the steps afterwards, such as, the intermediate storage and the creation of water-saving irrigation technologies. Although the technical processes of wastewater treatment for reuse in agricultural irrigation are more or less well known, there are many other factors that have not been well reflected yet. Some examples include unclear responsibilities, uncertainties about which water quality standards are to be applied, insufficient budgets, and a lack of trained operating personnel. These factors pose enormous challenges to the implementation of water reuse projects and their reliable and smooth operation. To ensure sustainability in water reuse projects, it is also essential to take many other aspects, including health, ecological, institutional, economic, and social aspects, into account.

The DWA Matrix presented in this manuscript offers an overview of the various possibilities for wastewater treatment for reuse purposes and is intended to be a fast and simple decision-making tool. Although it should not be considered as a perfect solution, the DWA Matrix can be applied in most cases to achieve the first rough estimate. It enables or eases the making of a sensible and well-founded decision, even when expert knowledge is not available to its fullest extent.

Acknowledgements Authors wish to offer there sincere gratitude to the DWA Working Group BIZ-11.4 for granting permission to use material developed by them in this manuscript. Authors also wish to thank leading members of the same Working Group, Prof. Peter Cornel and Dr. Tim Fuhrmann for their assistance with reviewing this manuscript.

Appendix A: Definition of Lines in Table 1

Note: Tables and explanations in Appendix A and B are direct extracts from the DWA publication *Treatments Steps for Water Reuse* (DWA 2008). There are 44 lines in the DWA Matrix. However, only 1–41 are directly applicable to the present publication. Lines 42–44 represent non-agricultural applications of water reuse.

Table 2 Line headings with assessment parameters

Aspect				Line
Health risk	Operating personnel water treatment facility			1
	Users of reutilized water			2
Economic efficiency	Investment costs	Floor space required		3
		Structural engineering		4
		Mechanical engineering		5
		E+MCR technology		6
	Operating costs	Personnel requirement/ costs		7
		Energy requirement/costs		8
		Disposal of residues		9
		Operating resources (precipitants etc.)		10
		Preventative maintenance costs		11
Effects on the environment through operation of the facility	CH_4 emission			12
	Odor nuisance			13
	Sounds/noisiness			14
	Aerosols			15
	Insects (worms, flies etc.)			16
Requirements on the operating personnel	Operability/operating expenditure			17
	Expenditure for preventative maintenance			18
	Required training of operating personnel			19
Plant technology	Degree of mechanisation			20
	Robustness			21
	Process stability			22
	Ability to influence the discharge quality operationally			23
	Discharge quality (treatment performance)	COD/BOD elimination		24
		SS reduction		25
		Nutrient elimination	Ammonium	26
			Nitrate	27
			Phosphorus	28
		Reduction of pathogens	Viruses	29
			Bacteria	30
			Protozoa	31
			Helminths	32
		Colour/Odour		33
		Residual turbidity		34
		Salting-up due to process		35
	Accumulation of residues			36
Irrigation technology	Root irrigation			37
	Trickling irrigation			38
	Sprinkler/Spray systems			39
	Flooding			40

(continued)

Table 2 (continued)

Aspect		Line
Types of use	Agricultural irrigation	41
	Non-potable water (toilet flushing)	42
	Urban uses (irrigation, water for fire protection)	43
	Forestry irrigation	44

A.1 Lines 1–2 "Health Risk"

The health risk associated with the operating personnel (of water treatment facilities) and the users of reused water are assessed qualitatively according to the following categories:

Category	Remarks
High	E.g., with the handling of "hazardous" chemicals
Medium	Disinfection is possibly required
Low	If employment takes place only during the pre-treatment step

A.2 Lines 3–6 "Economic Efficiency—Investment Costs"

Details on economic efficiency are of general and comparable nature. The categorization into low, medium, or high is only to allow a general comparative consideration of the process. These categories are determined and limits are set based on characteristic German values per capita (total number of inhabitants and population equivalents, PT):

Category	Remarks
High	Costs > 1000 €/PT and surface requirement >1 m^2/PT
Medium	Costs >600 to 1000 €/PT and surface requirement >0.3 to 1 m^2/PT
Low	Costs $\leq$600 €/PT and surface requirement $\leq$0.3 m^2/PT

Provision of concrete values is largely dispensed with, as these are often non-transferable. From the very beginning, the determination of investment and operating costs will be carried out attentively for each project, as economic efficiency is a decisive factor for the assessment. However, experience shows that costs can vary strongly, both from country to country and also from region to region within a country. Here, the following constraints are noted:

- Market conditions and the state of competition at the location/in the country,
- Detailed specifications of the selected technology,

- Relationship of structural engineering to mechanical engineering and/or equipment of the selected technology,
- Share of personnel costs in the investment and operating costs in countries with low wages,
- Availability and procurement costs of operating resources (energy, spare parts, expendable items, chemicals etc.),
- The need to have and/or mobilize highly qualified personnel for preventative maintenance and maintenance.

In the assessment matrix, investment costs have been divided into the areas surface requirement, structural engineering, mechanical engineering, and E+MCR (Electro-, Measurement-, Control- and Regulation technology). When numerically given, the surface requirement is specified in m^2/PT, as the basic price is extremely country-specific.

Fundamentally, for quantitative comparison, some treatment steps are designed according to load and others according to hydraulic capacity. Correspondingly, investment costs are normally set on the basis of either the number of inhabitants and population equivalents in €/PT or the hydraulic capacity in €/(m^3/h). A conversion is sensible to a limited extent only and possible only under the assumption of a specific wastewater discharge per number of inhabitants and population equivalents.

A.3 Lines 7–11 "Economic Efficiency—Operating Costs"

The general comments made about investment costs apply along the same lines for operating costs of the considered treatment processes, which are subdivided as follows:

- costs for personnel and/or personnel requirements,
- costs for energy and/or energy requirement,
- costs for the disposal of residues (presumably under German constraints),
- costs for operating resources, such as precipitants and flocculants or other chemicals,
- costs for preventative maintenance.

The numerical values refer to German conditions for newly erected facilities. The transferability to other countries, according to the comments on the investment costs, is not directly given.

For some processes the overall operating costs in euros per cubic meter (€/m^3) of treated water are given in accordance with the following categories:

Category	Remarks
High	Costs >0.4 €/m^3 and $\leq$0.8 €/m^3
Medium	Costs >0.06 to 0.4 €/m^3
Low	Costs $\leq$0.06 €/m^3

The energy requirement is given in kilowatt hours (kWh) per cubic meter of treated water. These values are largely universal and are thus directly transferable. The following categories are given for the energy requirement:

Category	Remarks
High	Energy requirement >0.02 kWh/m^3 and $\leq$0.2 kWh/m^3
Medium	Energy requirement >0.002 to 0.02 kWh/m^3
Low	Energy requirement $\leq$0.002 kWh/m^3

A.4 Lines 12–16 "Effects on the Environment through the Operation of the Facility"

Environmental loadings on the operation of the facilities for water treatment are assessed qualitatively, based on the following criteria:

- CH_4 emission (or emission of climate damaging gases),
- odour nuisance,
- sound/noisiness,
- aerosols,
- insects (worms, flies, mosquitos etc.).

Category	Remarks
High	High environmental loading
Medium	Medium environmental loading
Low	Low environmental loading

A.5 Lines 17–19 "Requirements on Operating Personnel through the Operation of the Facility"

The existing level of training of operating personnel, especially in many developing countries and emerging markets, represents a limiting factor for the selection of possible technologies for water treatment. In the assessment matrix the requirements on personnel, regarding a controlled operation, are assessed for each treatment process based on the following criteria:

- Operability and and/or operating expenditure,
- Preventative maintenance expenditure,
- Necessary training for operating personnel.

Category	Remarks
High	High requirements
Medium	Medium requirements
Low	Low requirements

A.6 Lines 20–36 "Plant Technology"

Under the umbrella term "plant technology" the technical details are gathered together about the respective processes, in particular on the treatment performance. In addition to numerical literature data, the qualitative assessment categories, given below, are used.

The quality of the treated water and/or the treatment performance is assessed based on the following wastewater parameters, in relation to the degree of elimination:

- COD and BOD_5 (organic carbon compounds),
- SS (filterable substances, solid matter, suspended solids),
- Nutrients (ammonium, nitrate, phosphorus),
- Pathogens (bacteria, viruses, protozoa, helminths).

In the matrix the degree of elimination is given in % or by the concentration in the treated water in mg/l; the reduction of pathogens is given in logarithmic steps (log-steps). The following categories are used:

Category	Remarks
High	Degree of elimination >70% or 4–6 log steps
Medium	Degree of elimination 30–70% or 2–3 log steps
Low	Degree of elimination <30% or up to 2 log steps
No influence	Degree of elimination <5%
Not relevant	E.g., if employed for post treatment only

Further parameters are drawn upon for qualitative description of the properties and condition of the treated water:

- Colour and odour,
- Residual turbidity,
- Salting-up of the water during the treatment.

Category	Remarks
High	The treated water shows a high (residual) colouring/odour/residual turbidity
Medium	The treated water shows a medium (residual) colouring/odour/residual turbidity
Low	The treated water shows a low (residual) colouring/odour/residual turbidity
No influence	–

Additional non-quantifiable parameters are drawn upon for the direct description of plant technology and qualitatively assessed in a comparative manner:

- Degree of mechanisation,
- Robustness,
- Process stability,
- Ability of influencing the discharge quality operationally.

Category	Remarks
High	Higher degree
Medium	More medium degree
Low	Lower degree

The accumulation of residues due to the treatment process is assessed as follows:

Category	Remarks
High	>80 to 110 l/(PT·a) dewatered sludge for disposal
Medium	>40 to 80 l/(PT·a) dewatered sludge for disposal
Low	Up to 40 l/(PT·a) dewatered sludge for disposal
No accumulation	–

A.7 Lines 37-40 "Irrigation Technology"

In the case of a utilization of wastewater as irrigation water, for each treatment process it is stated whether the treated water can be employed using the given irrigation technologies.

Generally, the solid matter concentration (e.g., expressed through the DS content) for irrigation facilities with very fine elements or spray nozzles (as in the case of root or trickling irrigation) has to be very small and, therefore, a filtration is recommended or is necessary.

For irrigation technologies, with which a development and distribution of fine droplets and aerosol particles occurs (e.g., through sprinkler systems), the treated water should additionally be disinfected in order to minimize health risks, e.g., for field workers and neighboring inhabitants.

Category	Remarks
Suitable	Possibly, however, limitations due to necessary filtration or disinfection
Less suitable	Requires filtration
Not suitable	–
Not relevant	E.g. if employment as pre-treatment only takes place

A.8 Lines 41–44 "Utilization Options"

These lines detail for each treatment process, in accordance with the following categories, whether the utilization of the treated water is possible and/or is worthy of recommendation for the respective purpose:

Category	Remarks
Recommended	–
Possible	–

(continued)

(continued)

Category	Remarks
Not recommended	–
Not possible	–

Appendix B: Assessment of the Treatment Technology

Note: Tables in Appendix B are direct extracts from the DWA publication on Treatments Steps for Water Reuse (DWA 2008). For download of the matrix for individual adaptation please contact the DWA costumer service (info@dwa.de).

The assessment of the treatment steps discussed is illustrated in this Appendix. Selection of the level (low, medium, and high) or the numerical values for each dimension was conducted based on different sources, which are numbered from 1-35 in the table below. The examples enclosed in the next few pages use theses reference numbers, immediately next to wherever they are applied. All 35 references are listed in a legend at the end of Appendix B.

Annex: Assessment matrix of treatment steps of water for reuse mechanical treatment

Aspect			Line no	Mechanical treatment									
				Screening						Sedimentation			
				With precipitation/ flocculation		Without precipitation/ flocculation		Micro-sieving 10 μm		With precipitation/ flocculation		Without flocculation	
Health risk	Operating personnel water treatment facility		1	High (handling of chemicals)	25	Medium	25	Low	27	High (handling of chemicals)	28	Medium	28
	Users of reused water		2	Low (only as pre-treatment stage)	25	Low (only as pre-treatment stage)	25	Low (disinfection necessary)	27	Low (only as pre-treatment stage)	28	Low (only as pre-treatment stage)	28
Economic efficiency	Investment costs	Surface requirement	3	Low	25	Low	25	Low	27	Low (0.04–0.06 m^2/ PT)	6	low(0.02–0.04 m^2/ PT)	6
		Structural engineering	4	Medium (400–1000 €/(m^3/ h) + flocculation)	2	Low (400–1000 €/(m^3/h))	2	Low	27	Medium (250–1000 €/PT settling tank + 1–80 €/PT precipitation)	3	medium (250–1000 €/PT for settling tank)	3
		Mechanical engineering	5	Low	25	Low	25	Medium	27	Low	34	Low	34
		E+MCR technology	6	Low	25	Low	25	Low	27	Low	34	Low	34
	Operating costs	Personnel requirement/ costs	7	Low	25	Low	25	Low	27	Low	34	Low	34
		Energy requirement/costs	8	Medium (0.0117–0.017 kWh/m^3)	27	Medium (0.009–0.013 kWh/m^3)	27	low	27	Low ($\sim$0.002 kWh/ m^3)	5	Low ($\sim$0.001 kWh/ m^3)	5
		Disposal of residues	9	High	25	Medium	25	Low	27	High	34	Medium	34
		Operating resources (precipitant etc.)	10	High	25	Low (no operating resources)	25	Low	27	High	34	Low (no operating resources)	34
		Preventative maintenance costs	11	Low	25	Low	25	Low	27	Low	34	Low	34

(continued)

(continued)

Aspect		Line no	Mechanical treatment									
			Screening						Sedimentation			
			With precipitation/ flocculation		Without precipitation/ flocculation		Micro-sieving 10 μm		With precipitation/ flocculation		Without flocculation	
Effects on the environment through operation of the facility	CH_4– Emission	12	None	25	None	25	None	27	Low (only with long sedimentation times slight methane formation through anaerobic degradation process possible)	30	Low (only with long sedimentation times slight methane formation through anaerobic degradation process possible)	30
	Odour nuisance	13	High	29	High	29	Low	27	Low	29	Medium	29
	Sounds/noisiness	14	Low	29	Low	29	Low	27	Low	29	Low	29
	Aerosols	15	Low	29	Low	29	Medium	27	Low	29	Low	29
	Insects (worms, flies, etc.)	16	High	29	High	29	Low	27	Medium	29	Low	29
Requirements on operating personnely	Operability/operational expenditure	17	Medium	31	Low	25	Medium	31	Medium	31	Low	31
	Preventative maintenance expenditure	18	Medium	31	Low	25	Medium	31	Medium	31	Low	31
	Required training for operating personnels	19	Medium	29	Low	29	Medium (trained personnel required)	27	Medium	29	Low	29

(continued)

(continued)

Aspect				Line no	Mechanical treatment									
					Screening						Sedimentation			
					With precipitation/ flocculation		Without precipitation/ flocculation		Micro-sieving 10 μm		With precipitation/ flocculation		Without flocculation	
Plant technology	Degree of mechanisation			20	Low/medium	25	Low	25	High	27	Medium	27	Low	27
	Robustness			21	High	25	High	25	Medium	27	Medium	27	High	27
	Process stability			22	High	25	High	25	Medium	27	High	27	High	27
	Ability to influence the discharge quality operationally			23	Medium	25	Low	31	Low	31	Medium	31	Low	31
	Discharge quality (treatment performance)	COD/BOD elimination		24	Medium (Maximum 60%)	25	Low (Maximum 25%)	25	Low (>10% or <60 mg/l)	27	Medium/high (55–75% COD; 45–80% BOD)	6	Medium (25–35% COD; 30–35% BOD)	6
		SS reduction		25	High (maximum 95%)	25	High (85%)	25	Medium (> 30% or < 10 mg/l)	27	Medium/high (60–90%)	6	Medium (55–65%)	6
		Nutrient elimination	Ammonium	26	Low (ca. 10%)	34	Low (ca. 10%)	34	Low	27	Low (<30%)	6	Low (<30%)	6
			Nitrate	27	No influence (0%)	25	No influence (0%)	25	Low	27	No influence (0%)	34	No influence (0%)	3
			Phosphorus	28	High	25	Low (<10%)	25	Low	27	High (75–90%)	6	Medium/low(< 35%)	6
		Reductions of pathogens	Viruses	29	Low	34	Low	34	No detail	27	Low (1–2 log steps)	1	Low (0–1 log steps)	1
			Bacteria	30	Low	34	Low	34	No detail	27	Low (1–2 log steps)	1	Low (0–1 log steps)	1
			Protozoa	31	Low	34	Low	34	No detail	27	Low (1–2 log steps)	1	Low (0–1 log steps)	1
			Helminths	32	Low	34	Low	34	No detail	27	Medium (1–3 log steps)	1	Low (0–<1 log steps)	1
		Colour/odour		33	No influence	25	No influence	25	No influence	27	Low (with longer sedimentation times odour through anaerobic degradation processes possible)	30	Low (with longer sedimentation times odour through anaerobic degradation processes possible)	30
		Residual turbidity		34	Low	25	Medium	25	Low	27	Low	34	Medium	34
		Salting up due to treatment		35	Medium (salting through precipitation chemicals)	25	No influence	25	No influence	27	High (salting through precipitation chemicals)	30	No influence	30

(continued)

(continued)

Aspect		Line no	Mechanical treatment									
			Screening						Sedimentation			
			With precipitation/ flocculation		Without precipitation/ flocculation		Micro-sieving 10 μm		With precipitation/ flocculation		Without flocculation	
	Accumulation of residues	36	Medium (country-specific; 15–70 l/(PT.a))		Medium (country-specific; 15–60 l/(PT.a))		Low		High (730–2500 l/(PT.a) unstabilised, liquid or 40–110 l/(PT.a) dewatered sludge)	6	Low (330–730 l/(PT.a) unstabilised, liquid or 15–40 l/(PT.a) dewatered sludge)	6
Irrigation technology	Root irrigation	37	Not suitable	25	Not suitable	25	Suitable	27	Not suitable	10	Not suitable	10
	Trickling irrigation	38	Not suitable	25	Not suitable	25	Suitable	27	Not suitable	10	Not suitable	10
	Sprinkler/spray systems	39	Suitable (requires disinfection)	25	Not suitable	25	Suitable	27	Suitable (requires disinfection)	10	Suitable (requires disinfection)	10
	Flooding	40	Suitable	25	Suitable	25	Suitable	27	Suitable	10	Suitable	10
Types of use	Agricultural irrigation	41	Possible	29	Not recommended	29	Recommended	27	Possible	29	Possible	29
	Non-potable water (e.g. toilet flushing)	42	Not recommended	25	Not possible	25	Possible	27	Not recommended	29	Not possible	29
	Urban uses (e.g. irrigation. water for fire-protection)	43	Not recommended	25	Not possible	25	Possible	27	Not recommended	29	Not possible	29
	Forestry irrigation	44	Possible	25	Possible	25	Recommended	27	Possible	29	Possible	29

Wastewater ponds, wastewater storage and treatment tanks

Aspect			Line no	Wastewater ponds						Wastewater storage and treatment tank	
				Aerated/aerobic with sedimentation pond		Unaerated/anoxic/anaerobic		Downstream polishing pond			
Health risk	Operating personnel water treatment facility		1	Low	26.33	Low	26.33	Low	26.33	Low	26.33
	Users of reused water		2	Medium (disinfection necessary)	26.33	Medium (disinfection necessary)	26.33	Medium (disinfection necessary)	26.33	Low (with long retention time)	26.33
Economic efficiency	Investment costs	Surface requirement	3	High (0.25–0.5 m^2/PT)	6	High (1.2–3.0 m^2/PT)	6	High (3.0–5.0 m^2/PT)	6	High	6
		Structural engineering	4	Low (300–1 000 €/PT)	26.33	Low (300–1000 €/PT)	26.33	Low (300–1000 €/ PT)	26.33	Medium	26.33
		Mechanical engineering	5	Low	2	Low	2	Low	26.33	Low	26.33
		E+MCR technology	6	Low	2	Low	2	Low	26.33	Low	26.33
	Operating costs	Personnel requirement/ costs	7	Low	4	Low	4	Low	34	Low	26.33
		Energy requirement/costs	8	Medium	33	Low	33	Low	33	Low	26.33
		Disposal of residues	9	Medium	26.33	Medium	26.33	Low	26.33	Low	26.33
		Operating resources (precipitant etc.)	10	Low (no operating resources)	26.33	Low (no operating resources)	26.33	Low (no operating resources)	26.33	Low (no operating resources)	26.33
		Preventative maintenance costs	11	Low	26.33	Low	26.33	Low	26.33	Low	26.33
Effects on the environment through operation of the facility	CH_4– Emission		12	Medium (methane formation in settling areas trough anaerobic degradation process)	26.33	High (considerable methane formation through anaerobic degradation process)	26.33	Low (possible methane formation through degradation of residual loads and sludge)	26.33	High (considerable methane production through anaerobic degradation processes)	26.33
	Odour nuisance		13	Low	26.33	High (dependent on operation)	26.33	Low	26.33	Low	26.33
	Sounds/ noisiness		14	Medium (dependent on aeration)	26.33	None	26	None	26	None	26
	Aerosols		15	Medium (dependent on aeration plant)	26.33	Low	26.33	Low	26.33	Low	26.33
	Insects (worms, flies, etc.)		16	High (mosquitos)	26.33	High (mosquitos)	26.33	High (mosquitos)	26.33	High (mosquitos)	26.33

(continued)

(continued)

Aspect				Line no	Wastewater ponds						Wastewater storage and treatment tank	
					Aerated/aerobic with sedimentation pond		Unaerated/anoxic/anaerobic		Downstream polishing pond			
Requirements on operating personnely	Operability/operational expenditure			17	Low	26.33	Low	26.33	Low	26.33	Low	26.33
	Preventative maintenance expenditure			18	Low	26.33	Low	26.33	Low	26.33	Low	26.33
	Required training for operating personnels			19	Low	26.33	Low	26.33	Low	26.33	Low	26.33
Plant technology	Degree of mechanisation			20	Low	26.33	Low	26.33	Low	26.33	Low	26.33
	Robustness			21	High	26.33	High	26.33	High	26.33	High	26.33
	Process stability			22	High	26.33	High	26.33	High	26.33	High	26.33
	Ability to influence the discharge quality operationally			23	Low	26.33	Low	26.33	Low	26.33	Low	26.33
	Discharge quality (treatment performance)	COD/BOD elimination		24	Medium/high (65–80% COD; 75–85% BOD)	6	Medium/high (65–80% COD; 75–85% BOD)	6	Low (reduction residual loads/ balancing of effluent peaks)	26.33	Low (reduction residual loads/ balancing of effluent peaks)	10
		SS reduction		25	High (70–80%)	6	High (70–80%)	6	Low (reduction residual loads/ balancing of effluent peaks)	26.33	Low (reduction residual loads/ balancing of effluent peaks)	10
		Nutrient elimination	Ammonium	26	Low (<30%)	6	Medium (<50%)	6	Low (reduction residual loads/ balancing of effluent peaks)	26.33	Low (reduction residual loads/ balancing of effluent peaks)	10
			Nitrate	27	Low (<30% N_{tot})	6	Medium (<60% N_{tot})	6	Low (reduction residual loads/ balancing of effluent peaks)	26.33	Low (reduction residual loads/ balancing of effluent peaks)	10
			Phosphorus	28	medium/low (<35%)	6	Medium/low (<35%)	6	Low (reduction residual loads/ balancing of effluent peaks)	26.33	Low (reduction residual loads/ balancing of effluent peaks)	10

(continued)

(continued)

Aspect				Line no	Wastewater ponds						Wastewater storage and treatment tank	
					Aerated/aerobic with sedimentation pond		Unaerated/anoxic/anaerobic		Downstream polishing pond			
		Reductions of pathogens	Viruses	29	Low (1–2 log steps, dependent on retention time)	1	High (1–4 log steps, dependent on retention time)	1	High (1–4 log steps, dependent on retention time)	1	High (1–4 log steps, dependent on retention time)	1
			Bacteria	30	Low (1–2 log steps, dependent on retention time)	1	High (1–6 log steps, dependent on retention time)	1	High (1–6 log steps, dependent on retention time)	1	High (1–6 log steps, dependent on retention time)	1
			Protozoa	31	Low (0–1 log steps, dependent on retention time)	1	High (1–4 log steps, dependent on retention time)	1	High (1–4 log steps, dependent on retention time)	1	High (1–4 log steps, dependent on retention time)	1
			Helminths	32	Medium (1–3 log steps, dependent on retention time)	1	Medium (1–3 log steps, dependent on retention time)	1	Medium (1–3 log steps, dependent on retention time)	1	Medium (1–3 log steps, dependent on retention time)	1
		Colour/odour		33	Medium (colouration due to algae and bacteria)	26.33	High (colouration through algae and bacteria/odour through anaerobic degradation processes)	26.33	Medium (colouration due to algae and bacteria)	26.33	Medium (colouration due to algae formation and bacteria)	26.33
		Residual turbidity		34	Medium	26.33	Medium	26.33	Medium	26	Low	26.33
		Salting up due to treatment		35	Medium (danger of salting uo through evaporation)	26.33	Medium (danger of salting uo through evaporation)	26.33	Medium (danger of salting uo through evaporation)	26.33	Medium (danger of salting uo through evaporation)	26.33
	Accumulation of residues			36	Medium (periodic sludge clearance)	26.33	Medium (periodic sludge clearance)	26.33	Low (periodic sludge clearance)	26.33	Low (periodic sludge clearance)	26.33
Irrigation technology	Root irrigation			37	Suitable (requires filtration)	10	Suitable (requires filtration)	10	Suitable (requires filtration)	10	Suitable (requires filtration)	10
	Trickling irrigation			38	Suitable (requires filtration)	10	Suitable (requires filtration)	10	Suitable (requires filtration)	10	Suitable (requires filtration)	10
	Sprinkler/spray systems			39	Less suitable (requires disinfection)	10	Suitable	10	Suitable	10	Suitable	10
	Flooding			40	Suitable	10	Suitable	10	Suitable	10	Suitable	10

(continued)

(continued)

Aspect		Line no	Wastewater ponds						Wastewater storage and treatment tank	
			Aerated/aerobic with sedimentation pond		Unaerated/anoxic/anaerobic		Downstream polishing pond			
Types of use	Agricultural irrigation	41	Possible	26.33	Possible	26.33	Possible	26.33	Possible	26.33
	Non-potable water (e.g. toilet flushing)	42	Not recommended	26.33	Not recommended	26.33	Not recommended	26.33	Not recommended	26.33
	Urban uses (e.g. irrigation, water for fire-protection)	43	Not recommended	26.33	Not recommended	26.33	Not recommended	26.33	Not recommended	26.33
	Forestry irrigation	44	Possible	26.33	Possible	26.33	Possible	26.33	Possible	26.33

UASB (Anaerobic upflow sludge blanket reactors), activated sludge processes, biological filters, reed beds

Aspect			Line no	UASB (Anaerobic upflow sludge blanket reactors)		Activated sludge process				Trickling filter		Helophyte treatment plants	
						C removal		Nutrient elimination					
Health risk	Operating personnel water treatment facility		1	Low	28	Low	28	High (handling of chemicals)	28	Low	28	Low	28
	Users of reused water		2	Low (only as pre-treatment stage)	28	Medium (disinfection required)	28	Medium (disinfection required)	28	Medium (disinfection required)	28	Medium (disinfection required)	28
Economic efficiency	Investment costs	Surface requirement	3	Low (0.03–0.1 m^2/PT)	6	Low (0.12–0.25 m^2/PT)	6	Low (0.12–0.25 m^2/PT)	6	Low (0.12–0.3 m^2/PT)	S	High (3–5 m^2/PT)	6
		Structural engineering	4	Medium	26	Medium (100–800 €/PT)	2	Medium (200–900 €/PT)	2	Medium (200–600 €/PT)	2	High (1000–2000 €/PT)	24
		Mechanical engineering	5	Medium	30	Medium (40–80 €/PT)	2	Medium (40–80 €/PT)	2	Low	2	Low	24
		E+MGR technology	6	Medium	30	High	2	High	2	Low	2	Low	24
	Operating costs	Personnel requirement/ costs	7	Low	30	Medium (5–10 €/(PT.a))	8	Medium (5–10 €/(PT.a))	8	Low	4.9	Low (50–130 €/ (PT.a))	24
		Energy requirement/costs	8	Low	30	High (~0.110 kWh/ m^3)	5	High (~0.190 kWh/ m^3)	5	Medium (~0.085 kWh/ m^3)	5		
		Disposal of residues	9	Low	30	Medium (10–20 €/(PT·a))	8	Medium (10–20 €/(PT·a)	8	Low	49		
		Operating resources (precipitant etc.)	10	Low (no operating resources)	30	Medium (1–2.5 €/(PT·a))	8	Medium (1–2.5 €/(PT·a))	8	Low	4.9		
		Preventative maintenance costs	11	Low	32	Medium (2.5–5 €/(PT·a))	8	Medium (2.5-5 €/(PT·al)	8	Low	4.9		

(continued)

(continued)

Aspect		Line no	UASB (Anaerobic upflow sludge blanket reactors)		Activated sludge process				Trickling filter		Helophyte treatment plants	
					C removal		Nutrient elimination					
Effects en the environment through operation of the facility	CH_4- Emission	12	High (the methane load dissolved in the treated water (the more the higher the temp.) evaporates)	30	None	30	None	30	None (only if air flow insufficient possible formation of anaerobic zones with methane development)	30	Low (formation of anaerobic zones with methane development possible)	26
	Odour nuisance	13	Low	30	Medium	29	low	29	medium	30	low	30
	Sounds/noisiness	14	Low	30	Medium/high (dependent on plant technology)	29	Medium/high (dependent on plant technology)	29	None	26	None	26
	Aerosols	15	Low	30	Low/high (dependent on plant technology)	29	Low/high (dependent on plant technology)	29	Low	30	Low	30
	Insects (worms flies, etc.)	16	Low	30	Low	29	Low	29	High	30	High	30
Requirements on operating personnely	Operability/operational expenditure	17	Medium	30	Medium	31	High	31	Medium	30	Low	30
	Preventative maintenance expenditure	18	Medium	30	Medium	31	High	31	Medium	30	Low (periodic cutting of the plants)	30
	Required training for operating personnels	19	Medium	30	Medium	29	High	29	Medium	30	Low	30
Plant technology	Degree of mechanisation	20	Low	27	High	27	High	27	Medium	30	Low	27
	Robustness	21	Low	27	High	27	High	27	High	27	Low/medium	27
	Process stability	22	Low	27	High	27	High	27	High	27	High	27
	Ability to influence the discharge quality operationally	23	Medium	30	High	30	High	30	Medium	30	Low	30

(continued)

(continued)

Aspect				Line no	UASB (Anaerobic upflow sludge blanket reactors)		Activated sludge process				Trickling filter		Helophyte treatment plants	
							C removal		Nutrient elimination					
	Discharge quality (treatment	COD/BOD elimination		24	Medium/high (50 to 85–95%)		High (80–90% COD; 85–93% BOD)	6	High (80–90% COD; 85–93% BOD)	6	High (70–80% COD; 80–83% BOD)	6	High (75–85% COD; 80–90% BOD)	6
		SS reduction		25	Medium/high (65–80%)	6	High (87–93%)	6	High (87–93%)	6	High (87–93%)	6	High (87–93%)	6
		Nutrient elimination	Ammonium	26	Medium (<50%)	6	Low (ca. 20%)	3	High (>80%)	6	Medium/high (50–85%)	6	Medium/high (40–98% with seasonal variations)	29
			Nitrate	27	Medium (<60% N_{tot})	6	No effect (0%)	3	High (ca. 80%)	34	Medium (<60% N_{tot})	6	Low (0–17%)	24
			Phosphorus	28	Medium/low (<35%)	6	Low (0% wo. precipitation)/ high (ca. 90% with precipitation)	3	Low/ (30% wo. Precipitation)/ high (ca. 90% with precipitation)	3	Medium/Low (<35%) (only with precipitation)	35	medium/high (30–95% depending on age)	29
		Reductions of pathogens	Viruses	29	Low (0–1 log steps)	1	Low (0–2 log steps)	1	Low (0–2 log steps)	1	Low (0–2 log steps)	1	Low (1–2 log steps)	1
			Bacteria	30	low (0.5–1.5 log steps)	1	Low (1–2 log steps)	1	Low (1–2 log steps)	1	Low (1–2 log steps)	1	Medium/low (0.5–3 log steps)	1
			Protozoa	31	Low/(0–1 log steps)	1	Low (0–1 log steps)	1	Low (0–1 log steps)	1	Low (0–1 log steps)	1	Low (0.5–2 log steps)	1
			Helminths	32	Low (0.5–1 log steps)	1	low (1–<2 log steps)	1	Low (1– <2 log steps)	1	Low (1–2 log steps)	1	Medium (1–3 log steps)	1
		Colour/odour		33	High (formation of odour substances due to anaerobic degradation)	30	Low (with correct operation)	30	Low (with correct operation)	30	Low (possible formation of odour substances under anaerobic conditions)	30	Low (possible formation of odour substances under anaerobic conditions)	30

(continued)

(continued)

Aspect		Line no	UASB (Anaerobic upflow sludge blanket reactors)		Activated sludge process: C removal		Activated sludge process: Nutrient elimination		Trickling filter		Helophyte treatment plants	
	Residual turbidity	34	Medium		Medium		Medium		Medium	30	Medium	30
	Salting up due to treatment	35	No effect	30	Low	30	Medium (salting up due to precipitation chemicals for P removal)	30	Low (danger of salting up through precipitant or water evaporation only with higher recirculation rate strong sunrays, lower air humidity)	30.34	low (danger of salting up through evapo-transpiration via the plants)	30
	Accumulation of residues	36	Low (70–220 l/(PT·a) unstabilised, liquid or 10–35 l/(PT·a) dewatered sludge)	8	High (1100–3000 l/(PT·a) unstabilised. liquid or 35–90l/(PT.a) dewatered sludge)	6	High (1100–3000 l/PT·a) unstabilised, liquid or 35–90 l/(PT.a) dewatered sludge)	6	Medium (360–1800 l/(PT·a) unstabilised. liquid sludge or 35–80 l/(PT·a) dewatered sludge)	6	Medium/high (plant cutting)	30
Irrigation technology	Root irrigation	37	Not relevant (pre-treatment only)	10	Suitable (requires filtration)	10	Suitable (requires filtration)	10	Less suitable (necessary filtration)	10	Less suitable (necessary filtration)	10
	Trickling irrigation	38	Not relevant (pre-treatment only)	10	Suitable (requires filtration)	10	Suitable (requires filtration)	10	Less suitable (necessary filtration)	10	Less suitable (necessary filtration)	10
	Sprinkle/spray systems	39	Not relevant (pre-treatment only)	10	Suitable (requires disinfection)	10	Suitable (requires disinfection)	10	Suitable (requires disinfection)	10	Suitable (requires disinfection)	10
	Flooding	40	Not relevant (pre-treatment only)	10	Suitable	10	Suitable	10	Suitable	10	Suitable	10
Types of use	Agricultural irrigation	41	Not recommended	30	Recommended	29	Recommended	20	Possible	30	Possible	30
	Non-potable water (e g flushing of toilets)	42	Not possible	30	Not recommended	29	Possible	29	Not recommended	30	Not recommended	30
	Urban uses (e.g. irrigation, water for fire-protection)	43	Not possible	30	Not recommended	29	Possible	29	Not recommended	30	Not recommended	30
	Forestry irrigation	44	Possible	30	Recommended	29	Recommended	29	Possible	30	Possible	30

Filtration (downstream), precipitation/flocculation (downstream), membrane technology

Aspect			Line no	Filtration (downstream)						Precipitation/ flocculation (downstream)		Membrane technology			
				Quick filtration (coarse)		Slow sand filtration		Double layer filtration				UF/MF		NF/RO	
Health risk	Operating personnel water treatment facility		1	Filtration (downstream)	28	Low	28	Low	28	High (handling of chemicals)	28	High (handling of chemicals)	28	High (handling of chemicals)	28
	Users of reused water		2	Medium (disinfection necessary)	28	Medium (disinfection necessary)	28	Medium (disinfection necessary)	28	Medium (disinfection necessary)	28	Low	28	Low	28
Economic efficiency	Investment costs	Surface requirement	3	Low	30	Low	30	Low	30	Low	30	Low	30	Low	30
		Structural engineering	4	Low (25–60 €/ PT)	11	Low (25–60 €/ PT)	11	Low	32	Low	32	High (4000–8000 €/(m^3/h))	12, 13, 14, 15	High	34
		Mechanical engineering	5					Low	34	Low	32			High	34
		E+MCR technology	6					Low	34	Low	32			High	34
	Operating costs	Personnel requirement/costs	7	Low	11	Low	11	Low	34	Low	32	Medium (0.26–0.4 €/m^3)	12, 13, 14, 15	High	34
		Energy requirement/costs	8	Low	33	Low	33	Low	33	Low ($\sim$0.001 kWh/m^3)	5			High (0.45–0.70 \$/m^3 desalination)	10
		Disposal of residues	9	Low	11	Low	11	Low	34	Medium	32			High	34
		Operating resources (precipitant etc.)	10	Low	11	Low	11	Low	34	Medium	32			High	34
		Preventative maintenance costs	11	Medium	11	Medium	11	Medium	34	Medium	32			High	32
Effects on the environment through operation of the facility	CH_4– Emission		12	None	30	None	30	None	30	None	30	None	30	None	30
	Odour nuisance		13	Low	27	Low	27	Low	27	Low	30	Low	30	Low	30
	Sounds/noisiness		14	Low	27	Low	27	Low	27	Low	30	Low	30	Low	30
	Aerosols		15	Low	27	Low	27	Low	27	Low	30	None	30	None	30
	Insects (worms, flies, etc.)		16	Medium	27	Medium	27	Medium	27	Low	30	None	30	None	30
Requirements on operating personnely	Operability/operational expenditure		17	Medium	31	Medium	31	Medium	31	Medium	30	High	30	High	30
	Preventative maintenance expenditure		18	High	31	High	31	High	31	Medium	30	High	30	High	30
	Required training for operating personnels		19	High (trained personnel necessary)	27	High (trained personnel necessary)	27	High (trained personnel necessary)	27	High (trained personnel necessary)	30	High (trained personnel necessary)	30	High (trained personnel necessary)	30

(continued)

(continued)

Aspect				Line no	Filtration (downstream)						Precipitation/ flocculation (downstream)		Membrane technology			
					Quick filtration (coarse)		Slow sand filtration		Double layer filtration				UF/MF		NF/RO	
Plant technology	Degree of mechanisation			20	Low	27	Medium	27	Medium	27	Low	27	High	27	High	27
	Robustness			21	Medium	27	Medium	27	High	27	High	27	Medium	26	Medium	27
	Process stability			22	High	27	High	27	High	27	High	27	High	27	High	27
	Ability to influence the discharge quality operationally			23	High	30	High	30	High	30	High	30	High	30	High	30
	Discharge quality (treatment performance)	COD/BOD elimination		24	Low (>20% or <40 mg/l)	11	Low (>20% or <40 mg/l)	11	Low (>20% or <40 mg/l)	11	Low	30	High (with aeration ca. 89–96% or COD <30 mg/l, BOD <5 mg/l)	12, 13, 14, 15	Not relevant (post-treatment only)	30
		SS reduction		25	Medium/high (>50% or <5 mg/l)	11	Medium/high (>50% or <5 mg/l)	11	Medium/high (>50% or <5 mg/l)	11	High	30	High (almost 100%)	12, 13, 14, 15	High	26
		Nutrient elimination	Ammonium	26	Medium (<5 mg/l)	11	Medium (<5 mg/l)	11	Medium (<5 mg/l)	11	Low (ca. 10%)	3	High (with aeration ca. 90% or 0.1–2 mg/l)	12, 13, 14, 15	Not relevant (post-treatment only)	30
			Nitrate	27	High (<10 mg/l)	11	High (<10 mg/l)	11	High (<10 mg/l)	11	No influence (0%)	3	Medium/high (4.5 mg/l)	12, 13, 14, 15	Not relevant (post-treatment only)	30
			Phosphorus	28	Medium (30% without flocculation)/ high (ca. 70% or <0.3 mg/l with flocculation)	11	Medium (30% without flocculation)/ high (ca. 70% or <0.3 mg/l with flocculation)	11	Medium (30% without flocculation)/ high (ca. 70% or <0.3 mg/l with flocculation)	11	High	3	High (with precipitation ca. 90% or 0.5–0.7 mg/l)	12, 13, 14, 15	Not relevant (post-treatment only)	30
		Reductions of pathogens	Viruses	29	Medium (1–3 log steps)	1	Medium (1–3 log steps)		Medium (1–3 log steps)	1	Medium (1–3 log steps)	1	High (2.5 – >6 log steps)	1	High (2.5– >6 log steps)	1
			Bacteria	30	Medium (0–3 log steps)	1	Medium (0–3 log steps)		Medium (0–3 log steps)	1	Low (0–1 log steps)	1	High (3.5– >6 log steps)	1	High (3.5– >6 log steps)	1

(continued)

(continued)

Aspect				Line no	Filtration (downstream)						Precipitation/ flocculation (downstream)		Membrane technology			
					Quick filtration (coarse)		Slow sand filtration		Double layer filtration				UF/MF		NF/RO	
			Protozoa	31	Medium (0–3 log steps)	1	Medium (0–3 log steps)		Medium (0–3 log steps)	1	Medium (1–3 log steps)	1	High (>6 log steps)	1	High (>6 log steps)	1
			Helminths	32	Medium (1–3 log steps)	1	Medium (1–3 log steps)		Medium (1–3 log steps)	1	Low (2 log steps)	1	High (>3 log steps)	1	High (>3 log steps)	1
		Colour/odour		33	No influence	30	No influence	30	No influence	30	No influence	30	No influence	30	No influence	30
		Residual turbidity		34	Low	11	Low	11	Low	11	Low	3	Low	34	Low	30
		Salting-updue to the treatment		35	No influence	30	No influence	30	No influence	30	Medium (salting-up due to precipitant chemicals)	30	Medium (salting-up due to precipitant chemicals)	34	No influence (but heavily salted concentrate for disposal)	30
	Accumulation of residues			36	Low	30	Low	30	Low	30	Low	30	Low (550–1100 l/(PT·a) stabilised, fluid or 17–34 l/(PT·a) dewatered sludge)	3	Medium (heavily salted concentrate for disposal)	30
Irrigation technology	Root irrigation			37	Suitable	10	Suitable	10	Suitable	10	Suitable	10	Suitable	10	Suitable	10
	Trickling irrigation			38	Suitable	10	Suitable	10	Suitable	10	Suitable	10	Suitable	10	Suitable	10
	Sprinkler/spray systems			39	Suitable	10	Suitable	10	Suitable	10	Suitable	10	Suitable	10	Suitable	10
	Flooding			40	Suitable	10	Suitable	10	Suitable	10	Suitable	10	Suitable	10	Suitable	10
Types of use	Agricultural irrigation			41	Recommended	27	Recommended	27	Recommended	27	Recommended	30	Recommended	30	Recommended	30
	Non-potable water (e.g. toilet flushing)			42	Possible	27	Possible	27	Possible	27	Possible	30	Recommended	30	Recommended	30
	Urban uses (e.g. irrigation, water for fire-protection)			43	Possible	27	Possible	27	Possible	27	Possible	30	Recommended	30	Recommended	30
	Forestry irrigation			44	Recommended	27	Recommended	27	Recommended	27	Recommended	30	Recommended	30	Recommended	30

Desinfection

Aspect			Line no	Disinfection											
				Membrane (UF)		UV		Ozone		Soil filter		Polishing pond		Chlorine	
Health risk	Operating personnel water treatment facility		1	High (handling of chemicals)	28	Medium	26	High (handling of chemicals)	28	Low	28	Low	28	High (handling of chemicals)	28
	Users of reused water		2	Low	28	Low	28	Low	28	Low	28	Medium (post-disinfection necessary)	26	Low (only with over-chlorination)	26
Economic efficiency	Investment costs	Surface requirement	3	Low	30	Low	30	Low	30	High	30	High	30	Low	30
		Structural engineering	4	High	34	Low (7–41 €/PT)	16	High (0.52 €/m^3)	17	High	18.19.20.21	Low	22.23	Low	34
		Mechanical engineering	5	High	34	Medium	26	High	32	Low	18.19.20.21	Low	22.23	Medium (Safety technology)	26
		E+MCR technology	6	High	34	Medium	26	High	17	Low	18,19,20,21	Low	22.23	Low	34
	Operating costs	Personnel requirement/costs	7	High (0.2–0.8 €/m^3)	7	Low (0.03–0.05 €/m^3)	7	Medium (0.05–0.2 €/m^3)	7	Low	18,19,20,21	Low	22.23	Low (0.04–0.06 €/m^3)	7
		Energy requirement/costs	8							Low	18,19,20,21	Low	22.23		
		Disposal of residues	9							Low	18,19,20,21	Low	22.23		
		Operating resources (precipitants etc.)	10							Low	18,19,20,21	Low	22.23		
		Preventative maintenance costs	11							Low	18,19,20,21	Low	22.23		
Effects on the environment due to operation of the facility	CH_4 emission		12	None	26	None	26	None	26	None	26	Small (possible methane formation with anaerobic degradation of residual loads and sludge)	30	None	26
	Odour nuisance		13	Low	30	Low	30	Low	30	Low	30	Low	30	Low	30
	Sound/noisiness		14	Low	30	None	26	Low	30	None	26	None	26	None	26
	Aerosols		15	None	30	None	30	None	30	Low	30	Low	30	None	30
	Insects(worms, flies etc.)		16	None	30	None	30	None	30	Medium	30	High (mosquitos)	30	None	30

(continued)

(continued)

Aspect				Line no	Disinfection											
					Membrane (UF)		UV		Ozone		Soil filter		Polishing pond		Chlorine	
Requirements son operating personnel	Operability/operating expenditure			17	High	30	Low	30	High	30	Low	30	Low	30	High	30
	Maintenance expense			18	High	30	Medium	26	High	30	Low	30	Low	30	High	30
	Necessary training of operating personnel			19	High (trained personnel required)	30	Medium	26	High (trained personnel required)	30	Low	30	Low	30	High (trained personnel required)	30
Plant technology	Degree of mechanisation			20	High	27	Medium	27	Medium	27	Low	27	Low	27	Low	27
	Robustness			21	Medium	27	High	27	Medium	27	Medium	26	Low/medium	27	Medium	26
	Process stability			22	High	27	High	27	High	27	High	27	Medium/high	26	High	27
	Ability to influence the discharge quality operationally			23	High	30	High	30	High	30	Low	30	Low	30	High	30
	Discharge quality (treatment performance)	COD/BOD elimination		24	Not relevant (for post treatment only)	30	No influence	34	Not relevant (for post treatment only)	30	High (ca. 85%)	18,19,20,21	Low (reduction residual loads/balancing of effluent peaks)	26	No influence	34
		SS reduction		25	High	26	No influence	34	Not relevant (for post treatment only)	30	High (ca. 90%)	18,19,20,21	Low (reduction residual loads/balancing of effluent peaks)	26	No influence	34
		Nutrient elimination	Ammonium	26	Not relevant (for post treatment only)	26	No influence	34	Not relevant (for post treatment only)	30	High (ca. 80%)	18,19,20,21	Low (reduction residual loads/balancing of effluent peaks)	26	No influence	34
			Nitrate	27	Not relevant (for post treatment only)	26	No influence	34	Not relevant (for post treatment only)	30	Low (10% unplanted)/ high (70% unplanted)	18,19,20,21	Low (reduction residual loads/balancing of effluent peaks)	26	No influence	34

(continued)

(continued)

Aspect			Line no	Disinfection											
				Membrane (UF)		UV		Ozone		Soil filter		Polishing pond		Chlorine	
		Phosphorus	28	Not relevant (for post treatment only)		No influence		Not relevant (for post treatment only)		Medium (ca. 30% unplanted)/high (ca. 80% unplanted) performance sinks however with operating time		Low (reduction residual loads/balancing of effluent peaks)	26	No influence	34
	Reduction of pathogens	Viruses	29	High (2.5– >6 log step)	1	Medium (1– >3 log steps)	1	High (3–6 log steps)	1	Medium/low (1.5–2.5 log steps)	18.19.20.21	High (1–4 log steps)	1	Medium (1–3 log steps)	1
		Bacteria	30	High (3.5– >6 log step)	1	High (2– >4 log steps)	1	High (2–6 log steps)	1	Medium/low (1.5–2.5 log steps)	18,19,20,21	High (1–6 log steps)	1	High (2–6 log steps)	1
		Protozoa	31	High (>6 log step)	1	High (>3 log steps)	1	Low (1–2 log steps)	1	Medium/low (1.5–2.5 log steps)	18,19,20,21	High (1–4 log steps)	1	Low (0–1.5 log steps)	1
		Helminths	32	High (>3 log step)	1	No influence	1	Low (0–2 log steps)	1	Medium	26	Medium (1–3 log steps)	1	Low (0– <1 log steps)	1
	Colour/odour		33	No influence	30	Low (decolouration possible)	30	Low (removal of colour and odour substances)	30	Medium (repossible formation of odour substances with anaerobic conditions)	30	Medium (possible colouration due to algae; odour formation with anaerobic conditions)	30	Medium (aggravation of odour and taste if residual chlorine contained in water)	30
	Residual turbidity		34	Low	34	No influence	34	No influence	34	Low	18,19,20,21	Medium	30	No influence	34
	Salting up due to treatment		35	No influence	30	No influence	30	No influence	30	No influence	30	Small (danger of salting up through water evaporation with longer retention times, stronger sunrays, larger water surface)	30	Low	26
Accumulation of residues			36	Low (concentrate for disposal)	30	None	30	None	30	Low	26	Low (periodic sludge clearance)	30	None	30

(continued)

(continued)

Aspect		Line no	Disinfection						
			Membrane (UF)	UV	Ozone	Soil filter	Polishing pond	Chlorine	
Irrigation technology	Root irrigation	37	Suitable						10
	Trickling irrigation	38	Suitable						10
	Sprinkler/spray systems	39	Suitable						10
	Flooding	40	Suitable						10
Type of use	Agricultural irrigation	41	Recommended						30
	Non-potable water (e.g. For flushing toilets)	42	Recommended						30
	Urban uses (e.g. irrigation, water for fire-protection)	43	Recommended						30
	Forestry irrigation	44	Recommended						30

Legend of information sources

No.	Source
1	WHO, 2006a
2	Günthert and Reicherter, 2001
3	ATV-DVWK, 2000
4	DWA-Landesverband [Federal State Association] Bayern, 2005
5	MURL, 1999
6	Von Sperling and Chernicharo, 2006
7	ATV, 1998
8	Grünebaum and Weyand, 1995
9	Lenz, 2004
10	Alcalde et al., 2004
11	Strohmeier, 1998
12	Wedi, 2005
13	Engelhardt, 2006
14	Günder, 2001
15	Frechen, 2006
16	Schleypen, 2005
17	Cornel, 2006
18	Laber, 2001
19	Novak, 2005
20	DWA, 2006
21	Lützner, 2002
22	IRC, 2004
23	Ruhrverband, 1992
24	Barjenbruch and Al Jiroudi, 2005
25	Working Group (joint assessment)
26	Tim Fuhrmann (personal assessment)
27	Hans Huber (personal assessment)
28	Volker Karl (personal assessment)
29	Roland Knitschky (personal assessment)
30	Alessandro Meda and Peter Cornel (personal assessment)
31	Hermann Orth (personal assessment)
32	Holger Scheer (personal assessment)
33	Florian Schmidtlein (personal assessment)
34	Christina Schwarz (personal assessment)
35	Martin Marggraff (personal assessment)

References

AQUAREC. (2006). *Water reuse system management manual, AQUAREC (Integrated Concepts for Reuse of Upgraded Wastewater)*. In D. Bixio & T. Wintgens (Eds.). Luxembourg: Office for Official Publications of the European Communities, ISBN 92-79-01934-1.

Asano, T. (2007). *Water reuse: Issues, technologies and applications* (1st ed.). McGraw-Hill, March 2007, ISBN: 978-0-07-145927-3.

Ayers, R. S., & Westcot, D. W. (1985). *Water quality for agriculture*. Rome: Food and Agriculture Organization of the United Nations.

CCR. (2015). *Regulations Related to Recycled Water—Titles 22 and 17 California Code of Regulations*. California, USA: State Water Resources Control Board.

DIN. (1999). "Irrigation—Hygienic concerns of irrigation water," DIN 19650: 1978-09. Germany: Beuth Verlag GmbH.

DWA. (2008). Treatment Steps for Water Reuse. DWA Topics, Editor: Deutsche Vereinigung für Wasserwirtschaft, Abwasser und Abfall e. V. Hennef, Germany: German Association for Water, Wastewater and Waste, DWA.

EMWIS. (2007). Annex B—Case studies, Nov. 2007, Ergebnisse der Arbeitsgruppe Abwasserwiederverwendung, Euro-Mediterranean Information System on Know-how in the Water Sector, www.emwis.net/topics/waterreuse.

Firmenich, E., Fuhrmann, T., Gramel, S., Kampe, P., & Weistroffer, K. (2013). *Planning, execution and operation of reuse-projects*, DWA Slide Presentation (Training Material), available at: Deutsche Vereinigung für Wasserwirtschaft, Abwasser und Abfall e. V. Hennef, Germany: German Association for Water, Wastewater and Waste, DWA.

Fuhrmann, T., Scheer, H., Cornel, P. Gramel, S., & Grieb A. (2012). Water reuse: Diverse questions in view of an internationally increasing relevance. In KA -Korrespondenz Abwasser, Abfall – International Special Edition 2012 (pp. 19–24). Hennef, Germany: DWA/GFA.

Hettiarachchi, H., & Ardakanian, R. (2016). *Environmental resource management and nexus approach: Managing water, soil, and waste in the context of global change*. Switzerland: Springer Nature. ISBN 978-3-319-28593.

ISO. (2015). *Guidelines for treated wastewater use for irrigation projects*. ISO Standards 16075-1, 2 and 3, ISO.

Jimenez, B., & Asano, T. (2008). Water reuse: An International survey, contrasts, issues and needs around the world. In B. Jimenez & T. Asano (Eds.). London: IWA Publishing, 2007, Planned publication date: 1.2.2008, ISBN: 1843390892.

Kompetenzzentrum Wasser Berlin. (2012). *Presentation on reuse and OXIMAR*, not published, 2012, cited in Fuhrmann et al. 2012.

Lazarova, V., Asano, T., Bahri, A., & Anderson, J. (2013). Milestones in water reuse—The best success stories, IWA Publishing. www.iwapublishing.com/books/9781780400075/milestones-water-reuse.

United Nations. (2003). Water for people, water for life. The United Nations World Water Development Report. Executive Summary, UNESCO/Division of Water Sciences, Paris, France. http://unesdoc.unesco.org/images/0012/001295/129556e.pdf.

USEPA. (2004). Guidelines for water reuse EPA/625/R-04/108, Produced by Camp Dresser & McKee Inc. for United States Environmental Protection Agency, Washington DC, USA.

USEPA. (2012). Guidelines for Water Reuse, USEPA/600/R-12/618, United States Environmental Protection Agency, Washington, USA.

WHO. (2006). *World Health Organization guidelines for the safe use of wastewater, excreta and grey-water*. Geneva: World Health Organization. www.who.int/water_sanitation_health/publications/gsuweg2/en/.

Policy and the Governance Framework for Wastewater Irrigation: Jordanian Experience

Maha Halalsheh and Ghada Kassab

Abstract Many countries in West Asia are facing challenges related to management of their scarce water resources. Challenges are exacerbated by population growth and climate change. In this context treated wastewater for agricultural purposes can become a significant component. Accordingly, this chapter aims to present Jordan's experience in the field of wastewater treatment and reuse with interest in the governing legal and institutional framework and its evolution. Jordan is considered one of the few water-stressed countries in the region that has become successful in managing the limited water resources. In addition, treated wastewater use in agriculture is a well-developed practice in Jordan for decades and was originally motivated by the severe water scarcity and the demand to provide the agricultural sector with non-conventional irrigation water. The chapter is arranged in sections where the challenges and the solutions adapted at regional level are introduced together with historical development of sanitation paradigms. The chapter then digs into the details of Jordan's experience in treated wastewater use and presents how the legal and institutional arrangements were formulated. Challenges and opportunities associated with treated wastewater use in Jordan are then presented and required improvements as recognized by the government are introduced.

Keywords Sanitation · Wastewater management · Treated wastewater
Agricultural irrigation · Institutional arrangement · Legal framework
Policy implementation

M. Halalsheh (✉)
Water, Energy and Environment Center, The University of Jordan, Amman, Jordan
e-mail: halalshe@ju.edu.jo; nameer_maha@hotmail.com

Present Address:
M. Halalsheh
GIZ, Amman, Jordan

G. Kassab
Civil Engineering Department, University of Jordan, Amman, Jordan

H. Hettiarachchi and R. Ardakanian (eds.), *Safe Use of Wastewater in Agriculture*,
https://doi.org/10.1007/978-3-319-74268-7_5

1 Wastewater Irrigation in West Asia

West Asia is facing major water challenges due to scarcity, growing population, urbanization, and other industrial/development needs. Coupled with the fragile arid environment and its low resilience in the face of different activities, decision makers are left with major responsibilities to achieve safe and dependable water and food supplies in the future. Fresh water scarcity means greater risks for a community's ability to grow and create jobs (AFED 2014). Likewise, current regional political unrests combined with increased stress on economy have exerted serious threats to sustainable development. The situation has resulted in two major governing priority themes, namely water-energy-food, and peace-security-environment nexuses for the region (UNEP 2016). However, such priorities should not be examined in isolation from social, economic and institutional priorities, if the scope of impact of suggested solutions is called to have long and lasting effects.

Climate change will also threaten water and food security in the region due to the projected decrease in available fresh water resources for agricultural and food production (Almazroui 2012). Climate models project changes in the region's temperature, rainfall and sea level, which will have impacts on both availability and use of water resources (Sipkin 2012). Projections suggest 20% decrease in rainfall in the region over the next 50 years, while 40% reduction was predicted for some locations according to most global climate models (Meslemani 2008). The climate risk index, that classifies countries according to their exposure to climate change risks, has classified Iraq as the fifth most vulnerable country in the world in terms of decreased water, availability of food, extreme temperature conditions, and associated health problems (GEO-6 2016). Recent droughts have aggravated water crisis in Iraq and many studies warn that the Tigris and Euphrates might dry up by 2040 (Rowling 2014). Coupled with poor water quality, these stresses have displaced people from their livelihoods to seek for access to better drinking water (Rowling 2014). Other counties in the region were also rated as highly vulnerable, while Yemen was rated as extremely vulnerable. Climate change impacts will lead not only to a reduction in the quantity of water resources, but also will have an impact on water quality and is expected to increase the variability and frequency of extreme events (Glass 2010). It is therefore necessary to prepare for, and respond appropriately to the potential negative impacts of climate change.

On the demand side, a reduced per capita water share was observed in many countries in the region, partly as a result of the recent increase in the cross-border influx of refugees. The political unrest has recently arisen in several countries, including Iraq, Syria and Yemen, which resulted in a direct impact on water supply and sanitation services. Overexploitation of groundwater resources throughout West Asia was also observed and has resulted in deterioration of water quality, seawater intrusion, depletion and salinization of aquifers, and rising pumping costs. Depletion of non-renewable groundwater has been, moreover, observed with the expansion of agriculture. An increase of about 82% in the region's total blue water withdrawals for agriculture, and domestic use between 2000 and 2012 was noticed. The agricultural sector in almost all countries is by far the largest consumer of water resources (Abuzeid 2014) leaving little amounts for domestic and industrial sectors. All aforementioned challenges called for urgent responses in order to reduce the gap between water supply and water demand.

"Integrated resources management" is one of the best approaches that can help us make the best use of water resources in an era of water scarcity and climate change. The approach entails coordinating land and water management, recognizing water quantity and water quality linkages, improving techniques to manage demand and conserve water and learning through adaptive management experiments. In this regard, reallocating water towards domestic and industrial sectors—rather than agriculture—may be a critical and provocative way to adjust to water scarcity and enhance water availability. Although sector water reallocation may not have been announced as a policy in many countries, the highest priority given to the domestic water use have resulted in water reallocation from the agricultural sector (Abuzeid and Elrawady 2014). For instance, Iraq, Jordan and Qatar have witnessed significant sector water reallocation. Furthermore, Jordan has established a stand-alone reallocation policy and a substitution policy in 2016. The trend of reallocating fresh water for domestic use and allocating non-conventional water, such as treated wastewater and agricultural drainage, to agriculture is likely to be part of future water management in the whole region (Abuzeid 2014). Potential volume of non-conventional water resources in West Asia is estimated at 1.27 billion cubic meters of treated wastewater (Abuzeid and Elrawady 2014). This is in addition to the other non-conventional resources such as agricultural drainage and desalinated brackish and sea water. Obviously, wastewater contributes as a renewable water resource for agricultural expansion (Abuzeid 2014).

Full valorization of wastewater in agriculture requires integrated planning and likewise is critical to meet countries' obligations to many Sustainable Development Goals (SDGs), particularly SDG6 on water and sanitation. Notwithstanding that many countries of the region have relatively a good match between collected and treated wastewater, there is still a high demand outside large cities and in newly urbanized areas to receive better services. Wastewater in such areas is still discharged directly into the environment and only partly used for irrigation purposes, though unsafely. In many cases, wastewater and excess irrigation water infiltrates to reach the groundwater causing nitrate and pathogenic contamination. For instance, elevated nitrate concentration and pathogenic contamination were both reported for some springs in north Jordan due to domestic wastewater leaking from upstream nonpoint sources, principally cesspools. Contamination had resulted in closure of some drinking water springs, while it exerted additional treatment burden in some other cases. Obviously, serving rural scattered communities and rapidly expanded urban areas is crucial to protect scarce water resources and provide non-conventional water source for agricultural irrigation.

2 History of the Sanitation Paradigms

Conventional sewerage network and centralized wastewater treatment options are so far the dominant sanitation paradigm. Notwithstanding that this conventional centralized wastewater management scheme is generally no option for small-scattered communities and rapidly expanded peri-urban areas, it should be noted that utilizing fresh water to flush excreta to a sewerage network is not the zenith of scientific achievements particularly in water scarce countries. This historical practice was re-initiated more than 150 years ago when very little was known about fundamentals of water physics and chemistry and when practically applied microbiology was still not discovered. Minimizing fatal diseases breakouts in the nineteenth century was the main concern, and hence, wastewater was shipped as far as possible away from communities utilizing existing Roman sewer networks found in major European cities. In fact, bad smell was blamed by that time to be the cause of diseases as presented to the British parliament by the Chair of the Health Board on 1849. The Chair Edwin Chadwick stated that miasma was the main cause of death and the decision was made to transport all sewage outside the Victorian city of London and discharge it in the Themes River. The concept spread in other European cities and this paradigm became dominant with time resulting in complete division between citizens-consumers at one-hand and service providers at the other hand. Sanitation services became invisible and comfortable at the consumers side and associated risks disappeared from the world-life within served communities. However, the financial burden associated with this paradigm had restricted service provision for majority of population around the globe. Currently, 60% of the global population is not provided with sanitation services (Rachel et al. 2013), while almost 80% of the collected wastewater is discharged to the environment without treatment. Apparently, wastewater shipping is not necessarily what would be done today if countries had the chance to start again. Current advanced understanding of chemistry, physics and microbiology of wastewater, which was gained during the previous century, coupled with some other factors like limited resources and energy costs encourage us to find wastewater management alternatives. One such attractive alternative is to link sanitation management to cities' economic development (Kone 2010) through resource conservation and recovery. This new sanitation paradigm brought wastewater into the forefront and made the invisible sanitation services visible again (van Vielt et al. 2010). Consequently, all recently proposed sanitation alternatives required high level of community (consumer) involvement. The new paradigm calls for decentralization and sustainability and proposes better management for the limited resources by taking into account different pillars of decentralized sustainable sanitation including stakeholders' participation, technical feasibility, economic feasibility and legal and institutional arrangements. The proposed paradigm can be best implemented in non-serviced areas be it prei-urban, rural, or otherwise. Main differences between "old" and "new" paradigms are shown in Table 1.

Table 1 Paradigm shifts addressing water and sanitation infrastructure (van Vielt et al. 2010)

Old paradigm	New paradigm
Slow implementation	Rapid implementation
Prescriptive technologies	Adaptive solutions
Low social acceptance criteria	High social acceptance criteria
One water quality type fits all	Provision of water quality based on use
Low priority on energy efficiency	High priority for energy efficiency
"Siloed" health, economic, engineering	Integrated systems approach
Financing via taxes, subsidies, tariffs	Innovative financing and business models
Centralized energy provider	Distributed energy systems
Less priority on resource conservation	High priority on resource conservation

Notwithstanding the substantial benefits of the new paradigm, it is still beyond the required implementation level due to many reasons including the discouraging institutional environment and the lack of enforcement. Currently, opposite to central wastewater management systems, wastewater in small communities is not usually managed by the government. In general, they depend on house on-site sanitation systems, consisting mainly of cesspools, which are handled by self-organized private stakeholder upon demand. For instance, septage accumulating in the cesspools in Jordan is either transferred to special treatment plants or, in the absence of proper control, directly illegally discharged into the environment. Moreover, and in many instances where law enforcement is weak, households do not find a necessity to discharge septage since wastewater infiltrates into the soil and cesspools would rarely become full to present a nuisance to the inhabitants. Household may find it more convenient to close the cesspool when it becomes full and create another one, particularly when land space is available.

The issues faced by the small communities and peri-urban areas in getting access to sustainable sanitation services are manifold. Main challenges are summarized as follows:

1. Diseconomy of scale of sewer networks in less densely populated areas render conventional (and sometimes non-conventional) wastewater collection systems not feasible.
2. Innovation challenges, in which the new sanitation paradigm is a multi-stakeholders approach, and requires high level of community (consumers) involvement. It should be noted that social acceptance is not generally achievable on a short-term basis and requires specialized and long-term customized awareness campaigns oriented and designed case by case.
3. Most governmental authorities do not plan or invest in non-conventional sanitation alternatives; for instance, proper fecal sludge management options. Obviously, by investing in fecal sludge management, authorities/utilities may end up treating lesser volumes of wastewater per capita, while avoiding investment required to provide sewer connections to all (Reymond et al. 2016). However, public sector often lacks capacities and incentives needed for proper

planning and management of wastewater generated by small communities. Additionally, low technology small-scale wastewater treatment plants or on-site treatment systems are not as noticeable as large-scale systems, which make the later more appealing to decision makers. Apparently, existing environments tend to encourage high technology and still follow the top-bottom approach that has been so far implemented in centralized wastewater treatment systems.
4. Many clusters in rural communities and prei-urban areas are informal. Such clusters are not recognized by authorities and hence, provisions for services are unthinkable.
5. Non-conventional sustainable sanitation services would require the development of different and lenient regulations as compared to centralized sanitation services in order to allow for sustainable business models. Consequently, different institutional arrangements might be required.

Addressing the above listed challenges as well as challenges related to sanitation services provision require us to create an adequately enabling environment in which regulations, institutional arrangements, and social acceptance are prioritized. Moreover, technical feasibility and economic feasibility are also main concerns. The Kingdom of Jordan presents a good example in West Asia region with respect to sanitation services provision and reclaimed water use. Jordan had made impressive progress with respect to creation of enabling environment for both conventional centralized and non-conventional decentralized sanitation services. Although experience in sustainable decentralized sanitation services is still limited, Jordan has stepped forward and developed its own policy framework for decentralized sustainable sanitation that are planned for communities with less than 5000 inhabitants. Obviously, the main motivation of Jordan for the development of such policy was groundwater protection in view of the very limited fresh water resources. Moreover, achieving SDG 6 and the consequent international obligation was another main motivation behind the development of such policy. The following sections will further present and discuss Jordan's experience with respect to the enabling environment created for sanitation services provisions and use of wastewater.

3 Current Status of Wastewater Irrigation in Jordan

Water sector in Jordan is characterized by water scarcity issues exacerbated by the increasing water demand due to high population growth and economic development needs (Ministry of Water and Irrigation 2016). Challenges related to high population growth have been recently aggravated by an influx of refugees particularly those resulted from the ongoing political unrest in the region with around 650,000 reported Syrian refugees and 750,000 Syrian residents. Furthermore, water scarcity challenges are exacerbated by climate change and the associated augmented drought conditions. In fact, the average per capita annual renewable water share does not exceed 100 m^3, which is far below the global threshold of severe water scarcity which is reported at 500 m^3/capita/year. Moreover, the competition among domestic, agriculture, and industrial sectors present a serious water sustainability challenge. Only 5% of land receives enough rainfall to support cultivation. While farmers irrigate less than 10% of the total agricultural land, agricultural water requirements represented around 60% of total national water needs, which is estimated to be 700 MCM (million m^3) while at the same time, agricultural sector contributed only 3–4% to GDP in 2013 (Ministry of Water and Irrigation 2016). In fact, Jordan's system of subsidies affects the use of irrigation water, which necessitates strict rationing to allocate the remaining water resources. Appropriate water pricing can be used for optimizing cropping patterns and water distribution, which can also substantially increase agricultural production (Olmstead 2014; Ministry of Water and Irrigation 2016). Different irrigation technologies have been adopted particularly drip irrigation, which has resulted in yield gains and water savings.

Notwithstanding the severe water shortage; Jordan is one of few countries in the world to have managed its freshwater resources relatively well. The country has 97% water network coverage; one of the highest coverages in the region. Moreover, Jordan is currently thriving to improve water availability by influencing water demand behavior, optimizing water transfer and allocations, reusing reclaimed water in irrigation, and providing additional fresh water source by desalination. The Government of Jordan (GOJ) has recently developed and adopted several policies in face of confronts associated with water shortage. Issued policies include substitution policy, reallocation policy, decentralized wastewater management policy, National Water Strategy 2016–2025, and climate change policy. The Ministry of Water and Irrigation (MWI) is currently developing action plans for such policies in order to optimize scarce water resources management.

Use of wastewater in agriculture is a well-established practice in Jordan since decades and has been identified as a priority as will be described later. The country has managed to provide 63% of its population (totaling 9 million inhabitants) with sewerage network. All collected wastewater is being treated in 31 wastewater treatment plants distributed all over the country. More than 90% of the treated wastewater is used mainly for agricultural production. The rest of the population is served by house onsite management systems consisting mainly of cesspools. The Government's strategy and emphasis on wastewater collection and treatment is relatively comprehensive: the 31 central wastewater treatment plants are expected to treat 240 MCM/year by 2025 contributing to around 16% of the total water budget. As a minimum secondary biological treatment is applied and about 70% of the collected wastewater goes beyond and receives tertiary treatment.

The necessary adaptation to climate change requires integrated water resources management approach positioning wastewater precisely in the water cycle. Keeping in mind that wastewater is the only sustainable and increasing water resource; wastewater should be optimally utilized as a resource. The political situation in the region, especially the influx of war refugees, has also increased the magnitude of the water and environmental challenges within Jordan. Not excluding the carrying capacity of Jordan, the country has to carefully study available options that guarantee certain living standards as a host community. Obviously, water and water related issues lie at core when considering human dignity and equity. Food security adds to the challenge, which makes a combination of water and food securities among the top priorities of the country. Integrated resources management strategies are expected to play significant roles especially wastewater use in agriculture.

Although Jordan is a pioneer in using reclaimed water for agricultural production, the country is still facing some challenges that can be categorized into two sets. The first set of challenges is about the demanded increase in wastewater collection and treatment. This also entails the lack of (economically viable) services available for scattered communities in rural areas and for rapidly expanding peri-urban areas. The lack of such services presents a real barrier against the full utilization of the wastewater and, perhaps more importantly, prevention of potential groundwater pollution. The unaffordable investment costs of the conventional wastewater collection systems hindered the expansion of sanitation services to such communities. The only foreseen solution would be the implementation of the new paradigm that presents decentralized sustainable sanitation options as the core approach. The second set of challenges are related to the policy and capacity aspects including lack of socio-cultural acceptance, absence of legal framework, and related institutional arrangements. Another challenge that is linked indirectly to the limited valorization of wastewater is the limited science-policy interface. Any technological advance (wastewater-related or otherwise) generally takes long time before practical adoption. Demonstration projects as well as high level of communication and coordination are required to boost the application of such new concepts.

Increasing sanitation coverage is expensive, and the proposed shift in water sector expenditures from water supply to sanitation in Jordan covering the period 2011–2013 is a significant step towards increasing coverage (Ministry of Water and Irrigation 2016). The Wastewater Master Plan published through the ISSP (2014) provides a snapshot of the sanitation and wastewater treatment in Jordan and justifies the investments needed for wastewater collection. The following sections present Jordan experience in terms of how the management practices and polices evolved and supported reclaimed water use for agricultural production.

4 Evolution of Wastewater Governance in Jordan: Policies, Laws and Institutional Arrangements

Prior to addressing the evolution of wastewater management laws, policies, and reuse standards, it would be helpful to present the governmental institutional arrangements in wastewater management in Jordan. The governmental entities which are directly or indirectly involved in the field of wastewater management and reuse in Jordan are as follows (ACWUA 2011):

- The Ministry of Water and Irrigation (MWI).
- The Water Authority of Jordan (WAJ) and the Jordan Valley Authority (JVA) which are incorporated within the MWI.
- The Ministry of Environment (MoE).
- The Ministry of Health (MoH).
- The Ministry of Agriculture (MoA).
- Jordan Standards and Metrology Organization (JSMO)
- Jordan Food and Drug Administration (JFDA).

4.1 Polices Related to Wastewater Management and Use

Jordan adopted its first official wastewater use policy in 1978 (Haddadin and Shteiwi 2006). Wherein, wastewater was to be collected from the municipal sector and treated in wastewater treatment plants (WWTPs) to an acceptable degree. Reclaimed water then flows to King Talal Dam -the biggest dam- where it would be diluted with freshwater and the mixed water would progress from the dam to the Jordan Valley to be used for irrigation (Ghneim 2010). This policy was established to compensate Jordan Valley farmers for the amounts of fresh water pumped from the valley to the Capital Amman to meet the increasing demand on municipal water.

In 1998, a new policy called the 'Wastewater Management Policy' was approved by the Cabinet (Ghneim 2010). This policy had been the official governmental policy dealing with wastewater management and reuse between 1998 and 2009. Many important affirmations (Nazzal et al. 2000) were stated in that policy such as:

1. Wastewater shall be considered as a part of the Jordanian water budget.
2. The major towns and cities in Jordan should have adequate systems for wastewater collection and treatment in order to protect public health and the environment.
3. The priority of use should be assigned to irrigation.
4. The quality of the treated effluent should be monitored and the users must be alerted to any emergency which causes deterioration in the effluent quality so that they do not use the water unless remedial actions are taken.
5. Crops to be irrigated with reclaimed water or a mixture of reclaimed water and freshwater shall be chosen to accommodate the irrigation water, type of the soil and its chemistry, and reuse economics.
6. Crops irrigated with reclaimed water or mixed water should be monitored.
7. Sludge that results from wastewater treatment processes would be processed so that it could be used as a soil conditioner and a fertilizer. Care shall be practiced in order to comply with the regulations concerning the protection of public health and the environment.

8. Utilization of reclaimed and recycled water for industrial purposes shall be promoted.

Jordan's Water Strategy for 2008–2022 which is titled as "Water for Life" (MWI 2009) dedicated a separate chapter to wastewater. Several goals were set in this strategy, including:

1. Public health and environment shall be protected from all pollutants especially in the peripheries of WWTPs;
2. Treated wastewater shall comply with national standards and monitored in a periodic manner; and
3. The operation of all WWTPs shall be in accordance with international standards and manpower shall be trained in a way that ensures adequate operation.

Approaches were specified in the strategy in order to achieve the goals related to wastewater by 2022. Some of the key approaches are listed below:

1. An environmental impact assessment for each sanitation project shall be done. Any project of this sort shall not be executed unless it has been ascertained that there will not be any negative environmental impact as a result of its execution.
2. The process of wastewater treatment will be directed to the production of water fit for reuse in irrigation according to the WHO and Food and Agriculture Organization (FAO) Guidelines as a minimum. The use of treated wastewater for other purposes shall be subject to appropriate specifications.
3. Regular monitoring of treated wastewater quality will be performed at each WWTP.
4. Farmers will be encouraged to use modern and efficient irrigation technologies. Proper procedures shall be taken to protect the health of farm workers and prevent the contamination of crops with treated wastewater.
5. Public awareness about the danger of exposure to untreated wastewater and the significant value of treated wastewater for different end uses will be raised using different methods.
6. Public and farmers awareness programs will be designed and executed to encourage the use of treated wastewater and provide information about irrigation methods and produce handling. Such programs will be focused on ways to protect the farmers' health and the surrounding environment.

The current water strategy (2016–2025) is focused on the wastewater treatment and use as a component within integrated water resources management as mentioned earlier. Jordan will gradually substitute fresh water use in irrigation with wastewater wherever feasible. Water and wastewater pricing will be reconsidered according to water allocation models. This is of utmost importance if cost recovery is targeted. Moreover, centralized and decentralized wastewater treatment and reuse will be enhanced with special focus on centralized systems to serve for larger agricultural use projects. The strategy also does not encourage wastewater collection systems for communities with population not exceeding 5000 inhabitants. Such communities represent almost 28% of the total population in Jordan. Apparently, such communities can be served with other sustainable wastewater management according to existing needs.

4.2 Laws Related to Wastewater Management and Use

The Municipality Law No. 29/1955, which was introduced in 1955, was the first law related to wastewater management in Jordan (Ghneim 2010). Under this law, the governmental authorities of Amman, the capital of Jordan, and other municipalities were made responsible for construction, operation, and management of sewers (Ghneim 2010; ACWUA 2011). The Buildings Rural and Urban Planning Law No. 79/1966 was adopted by the government of Jordan in 1966 (Nazzal et al. 2000). This law enabled governmental agencies to regulate the disposition, collection, and discharge of wastewater which might cause inconvenience or damage (Nazzal et al. 2000).

In 1971, the Public Health Law No. 21 that provided a public health framework for the control of wastewater was enacted (Nazzal et al. 2000). According to this law, the MoH was granted the authority to regulate and monitor the quality of the treated wastewater. The Jordan valley Authority (JVA), which was briefly introduced above, was established in 1977 with the introduction of law No. 18/1977. Under this law, the role of planning and implementing infrastructure projects in the Jordan Valley was assigned to the JVA. Thus, the JVA presided over the construction and management of wastewater systems in the Jordan Valley (Nazzal et al. 2000).

The Martial law No. 2 was enacted in 1982 in order to deal with repercussions caused by the industrial sector, which was rapidly growing. The control of industrial wastewater discharges into natural water systems, such as Amman-Zarqa Basin in particular, was the main focus of this law (Nazzal et al. 2000). Later on, the Water Authority of Jordan (WAJ) was founded in 1983 according to the temporary Law No. 34/1983. WAJ responsibilities and duties were later defined by the Water Authority Law No. 18/1988, which stated that WAJ is in charge of implementing policies related to the provision of domestic and municipal water and wastewater disposal services. Its responsibilities include the design, construction, and operation of these services, as well as supervising and regulating the construction of public and private wells, licensing well-drilling rigs and drillers, and issuing permits to engineers and licensed professionals to perform water and wastewater-related activities (ISSP 2012; ACUWA 2011). The WAJ law was amended in 2001. Article 28 was introduced to allow for private sector participation in water and wastewater service delivery through the assignment of any of WAJ's duties or projects to any other body from the public or private sector or to a company owned totally or partially by WAJ. This amendment enabled WAJ to corporatize utilities and enter into build-operate-transfer (BOT) contract arrangements and other PSP options (ISSP 2012).

The year 1988 also witnessed the enaction of the Jordan Valley Development Law No. 19/1988. This law and its amendments state that it is not permitted to contaminate the Jordan Valley water or cause its contamination by introducing any material from any source to the valley. This law mandated JVA to undertake all works related to the development, utilization, protection and conservation of the water resources in the Jordan Valley. JVA's other responsibilities include (ISSP 2012):

1. Raising the efficiency of agricultural water use;
2. Studying, designing, implementing, operating and maintaining irrigation projects, all major dams and water harvesting structures; and
3. Defending Jordan's rights to trans-boundary waters.

In 1992, the Ministry of Water and Irrigation (MWI) was formed according to the by-law No. 54/1992 so as to merge water resources management in Jordan under one organization (Nazzal et al. 2000). The regulation of wastewater treatment and reuse was amidst the duties of the MWI (Nazzal et al. 2000).

In 1996, the Ministry of Health (MoH) discerned that water flowing to King Talal Dam (KTD) was polluted with treated wastewater discharged from As Samra WWTP and suspected that vegetables irrigated with this water could also be contaminated. It was also discerned to the MoH that these vegetables can be harmful to the health of those who consume it. Consequently, these vegetables became a health hazard according to the definition stated in the Public Health Law which required their destruction and taking the necessary procedures to prevent their transport to locations where they might be consumed. Consequently, the minister of health approved the destruction of all vegetables irrigated with water flowing in the Zarqa River -leading to KTD- within the aforementioned limits and also prohibited the use of Zarqa River water in any further irrigation of all types of vegetables until further notice. Based on this decision, use of the Zarqa River Water was limited to the irrigation of fodders, field crops, and trees on the condition of ceasing irrigation two weeks prior to the harvest.

In 2002, the Agriculture Law No. 44/2002 was issued. According to article 3A of this law, the Ministry of Agriculture (MoA) organizing and developing the agricultural sector in cooperation with the relevant authorities, whenever such cooperation is called for, is the responsibility of the MoA. This is to achieve several objectives such as the sustainability of utilizing natural agricultural resources without damaging the environment and the provision of protection for the environment, livestock, and plants. Article 3B of the same law states that the MoA is to accomplish the objectives of offering basic agricultural services in areas where the private sector either does not provide such services or provides them with a lack of competency and efficiency. Such services include performing laboratory analyses in domains related to agricultural production. Studies, research, and observations related to soil salinity are among the activities related to this law.

Article 15C of the Agriculture Law No. 44/2002 stated that the minister of agriculture issues the regulations which specify the conditions according to which treated wastewater, saline water, and brackish water can be used in the irrigation of crops. The minister specifies by means of these regulations the types of crops which are allowed to be irrigated by each of the aforementioned types of water. According to article 15D of this law, the use of wastewater or treated wastewater for the purpose of washing plants and agricultural products is prohibited. Anyone who violates article 15D is penalized with a fine of 100 Jordanian Dinars for each ton that has been washed with wastewater or treated wastewater and the violator is also required to destroy these plants and products.

Article 15E of the Agriculture Law No. 44/2002 states that anyone who uses wastewater or treated wastewater for the irrigation of crops in violation of the regulations issued according to paragraph C of article 15 is penalized with a fine of 50 Jordanian dinars for each Donum (= 1000 m^2) or part of it that was irrigated, and the violator is required to remove the planted crops and destroy them under the supervision of the MoA. In the event that the violator refuses or delays the removal and destruction of crops, the administrative governor has to order the destruction of crops on the expense of the violator and under the supervision of the MoA.

In 2006, the Environment Protection Law No. 52/2006 was issued. Article 4 of this law states that in order to achieve the objectives of environment protection and improvement of its various elements in a sustainable manner, the Ministry of Environment (MoE) in cooperation and coordination with the relevant authorities handles several tasks such as:

1. Monitoring, measurement, and follow up of the elements and components of the environment through specific centers accredited and certified by the MoE according to the adopted standards.
2. Issuance of the necessary environmental regulations for the protection of the environment and its elements, the conditions according to which agricultural projects can be established and related services which must be abided by and included in the prior conditions to issue or renew permits for these projects according to the stated legal standards.

In 2008, the Public Health Law No. 47/2008 was issued. Article 18B of this law stated that in the event of a disease outbreak or the occurrence of infections with this disease, the MoH has to take the necessary measures to prevent spreading of the disease such as monitoring public and private water resources, planted crops, and foodstuffs or other sources that may form possible means of carrying the infection. It is specified in article 21A of the same law that in order to prevent the outbreak of a disease which may result from wastewater, senior staff members from the MoH have the right to commission the authorities responsible of sanitation to take the necessary procedures to protect the public health during the time period specified by the former.

Article 51A of the Public Health Law No. 47/2008 states that the MoH, in coordination with the relevant authorities, handles the monitoring of wastewater, sewer networks, interior plumbing, and WWTPs according to its own legislations to ensure their compliance with health conditions. The MoH also has the responsibility of taking the appropriate procedures to prevent any damage to public health.

Article 51B of the Public Health Law No. 47/2008 states that if the MoH discerned that wastewater, sewer networks, plumbing, or WWTPs pose or may pose a threat to public health, then the ministry has to take all the necessary procedures to prevent the occurrence of the anticipated health risk.

In 2008, the Food and Drug Administration Law No. 41/2008 was issued. According to article 5 of this law, the Jordan Food and Drug Administration (JFDA) handles the task of monitoring the quality and validity of food in accordance with the adopted technical rules and legislations.

4.3 Standards Related to Wastewater Management and Use

Water quality laws, treated wastewater regulations, standards for treated wastewater discharge into streams and water bodies, and standards for reclaimed water use in irrigation that are currently enforced in Jordan, have been drafted based on the principles and regulations set by the WHO or on the more strict principles established by the State of California in the United States (Ulimat 2012). The national organization in charge of issuing such standards in Jordan is the Jordan Institute for Standards and Metrology (JSMO).

The reuse of wastewater for agricultural irrigation in Jordan was initially carried out according to the Health Guidelines for the Use of Wastewater in Agriculture and Aquaculture established by the WHO in 1989 (Ghneim 2010). The use of the 1989 WHO guidelines continued until the first Jordanian wastewater use standards were adopted in 1995. The Jordanian Standards JS 893/1995 was established by the WAJ and was approved by a technical committee for water and wastewater at JSMO (ACWUA 2011). The direct use of reclaimed water for irrigation of vegetable crops eaten raw such as cucumber, tomato, and lettuce was forbidden under the Jordanian Standards JS 893/1995 (McCornick et al. 2004). Sprinkler irrigation as well as the irrigation of crops during a period of 14 days prior to harvest were also forbidden (McCornick et al. 2004). Standards for the discharge of reclaimed water into wadis (streams) and water bodies, aquaculture, and groundwater recharge were addressed in the Jordanian Standards JS 893/1995 as well.

The Jordanian Standards JS 893/1995 were replaced by the Reclaimed Domestic Wastewater Standards JS 893/2002 (ACWUA 2011). The reasons for why the Jordanian Standards JS 893/1995 were amended and revised in 2002 can be summarized as follows:

1. The reuse activities covered within the Jordanian Standards JS 893/1995 needed to be expanded (ACWUA 2011).
2. Jordanian vegetables and fruits export market was hampered by the new tough regulations introduced by some import countries such as the Gulf Countries enforced prohibition on the process of importation (McCornick et al. 2004). Consequently, it became necessary to develop new standards which would ensure enhanced safety to both farmers and consumers (Ghneim 2010).

The Jordanian Standards JS 893/2002 was divided into two main groups which were the standards and guidelines. The Jordanian Standards JS 893/2002 also addressed groundwater recharge and the discharge of reclaimed water into streams, wadis, and areas of water storage. There were three categories of irrigation in the Jordanian Standards JS 893/2002. These categories were termed A, B, and C. Category A stood for the irrigation of vegetables eaten cooked, parking areas, sides of roads inside cities, and playgrounds. Category B referred to the irrigation of plenteous trees, green areas, and roadsides outside the cities. Category C referred to the irrigation of industrial crops, field crops, and forestry. Similar to the Jordanian Standards JS 983/1995, the direct use of wastewater in irrigation for vegetables eaten raw was also prohibited in the JS 893/2002 (MEDAWARE 2005). Use of wastewater through sprinkler irrigation was only allowed for golf courses and limited to night time. In that case, the sprinklers must not be accessible for use throughout the day and they must be of the movable type (MEDAWARE 2005). Same as in JS 893/1995, irrigation must be ceased two weeks before the harvest when reclaimed water is used for the irrigation of fruit trees.

The current standards governing the wastewater use in Jordan were introduced in 2006 (ACWUA 2011). The current standards—Jordanian Standards JS 893/2006, also include two main groups which are the standards group and the guidelines group. The standards group includes those standards with which the effluent produced by WWTPs must comply (Ulimat 2012). The guidelines group, on the other hand, is only taken for guidance purposes and if the values specified by the guidelines are exceeded, the end user must conduct studies in order to verify the impact of the produced effluent on public health and the environment (Ulimat 2012). The study must include suggestions on how the damage to public health or the environment can be prevented (Ulimat 2012).

The Jordanian Standards JS 893/2006 also addresses the discharge of reclaimed water into streams and water bodies, groundwater recharge, and irrigation. Similar to the JS 893/2002, there are three categories for irrigation termed A, B, and C. However, the JS 893/2006 also has an additional irrigation category which is the irrigation of cut flowers. The same principles regarding the direct reuse of reclaimed water for the irrigation of vegetables eaten raw, sprinkler irrigation, and the irrigation of fruit trees in the Jordanian Standards JS 893/2002 applied in the Jordanian Standards JS 893/2006.

According to the quality monitoring component in the Jordanian Standards JS 893/2006, the entity which owns the WWTP and the regulatory entity must make sure that the quality of the treated effluent conforms to the standards corresponding to its end use (Ulimat 2012). Laboratory tests must be performed by both the monitoring and operating entities according to the sampling frequency specified in the Jordanian Standards JS 893/2006 (Ulimat 2012). As for the evaluation component of the Jordanian standards JS 893/2006, it is specified that if any tested value doesn't comply with the standards stated for the discharge of the treated effluent into streams and water bodies, then an extra-confirmatory sample must be collected (ACWUA 2011). If the two samples exceed the limits allowed by the standards, then the concerned party will be informed so as to carry out correction measures as soon as possible (ACWUA 2011).

Recently, Jordanian standards JS 1766/2014 was issued as a guideline (non obligatory) determining usage of irrigation water including treated wastewater and taking into account WHO guidelines (2006). The issued guidelines deal with irrigation water in general regardless of water source. In the presented guidelines, level of crop restriction is determined by both irrigation water quality and irrigation system. It also includes a section which can be used as a guideline for selecting crops to be irrigated with different water qualities based on salinity. The latter is a major concern in irrigation water, especially in the Jordan valley.

In Jordan, the currently adopted program for monitoring crops irrigated with reclaimed water is based on several international standards (ACWUA 2011). These standards define methods need to be followed for sample collection, preparation, and analysis. The most important standards are (ACWUA 2011):

1. Standard for Sampling Fresh Fruits and Vegetables No. 1239/1999.
2. Fruits, Vegetables, and Derived Products—Decomposition of Organic Matter prior to Analysis—Wet method, Standard No. 1246/1999.
3. Fruits, Vegetables, and Derived Products—Decomposition of Organic Matter prior to Analysis—Ashing Method, Standard No. 1247/1999.

5 Policy Implementation and the Impact

Considerable achievements have been made in Jordan thus far with respect to developing comprehensive wastewater management policies and standards. Practically, reuse directorate at the WAJ is responsible for managing the process of wastewater use. Farmers planning to use treated wastewater have to apply at the directorate. Based on the area, reuse directorate establish an agreement with farmers and allocate a certain amount of water. The agreement is so far granted only for the cultivation of fodders and/or fruit trees. Water meters and valves are installed within the wastewater treatment plant and controlled by the WAJ staff. Treated effluent is carried by water lines to the adjacent farms and is being used directly for irrigation.

It is worth mentioning that although Jordanian regulations and standards (viz. regulations and conditions for the use of treated wastewater, brackish and saline water—issued by minister of agriculture based on agriculture law no 44/2002, article 15C and Jordanian standards 893/2006) allow irrigating vegetables eaten cooked with treated wastewater (i.e. direct use scheme for vegetables eaten cooked), WAJ has so far limited the direct use of treated wastewater to fodder crops, olive trees and forests trees. Financial returns would be significantly higher if farmers are granted licenses within the limits of the current standards to irrigate vegetables as well (Majdalawi 2003).

5.1 The Challenges

A national water reuse coordination committee (NWRCC) was formed as per the cabinet letter 57/11/6826 dated 21/5/2003 under the supervision of the secretary general of the WAJ. Other members of the committee represented the Royal Court, Ministry of Environment, Ministry of Health, Ministry of agriculture, Jordan Valley Authority, National Center for Agricultural Research and Technology Transfer, Royal Scientific Society, Farmers Union, Universities and Private sector. The main task of the committee is to coordinate with the reuse directorate (previously known as wastewater reuse unit) in order to eliminate overlapping between the ministries. However, the committee was not active and hardly any improvement was noticed.

As mentioned earlier, violation of the regulations related to the reuse of treated wastewater is met with the destruction of the crops in question together with a fine. Nevertheless, the Jordanian standards for wastewater use are not being fully implemented. Despite the fact that there are monitoring programs put in place to ensure the compliance with the regulations in terms of the water quality and the type of irrigated crops, farmers do not always conform to these conditions.

The lack of implementation of the standards can be attributed to several challenges such as (Ghneim 2010):

1. Certain treatment plants are currently being overloaded. Thus, the quality of the produced effluent doesn't always conform to the Jordanian Standards JS 893/2006. Currently many of these wastewater treatment plants are under upgrading processes;
2. The Jordanian Standards for the discharge and use of treated wastewater are relatively stringent. As a consequence, they are not always met;
3. The fact that there is relatively a large number of stakeholders involved such as the MoA, MWI, JVA, WAJ, MoE, and MoH may have caused an overlap of responsibilities and a lack of coordination (ACWUA 2011). A clear coordination set up does not exist, which result in loss of resources in view of multiplication of some tasks that exist between different stakeholders;
4. Some farmers use reclaimed water discharged into streams for unrestricted irrigation prior to the process of mixing the treated effluent with freshwater in reservoirs. This irrigation practice is considered illegal and violates the Jordanian Standards JS 893/2006. Not mentioning lack of irrigation water source, farmers probably follow this practice due to the lack of knowledge on their behalf;
5. The lack of financial resources can impede the rigorous monitoring intended to discover certain violations; and
6. The reuse of treated wastewater for irrigation is subject to competition from fresh water sources such as groundwater even though fresh water resources are scarce. This is due to the fact that the fees for using freshwater in irrigation are low and thus, farmers who happen to have access to freshwater have no incentive for using reclaimed water.

In addition to what has been discussed, there are some issues related to optimum utilization of quantity of reclaimed water rather than its quality. Firstly, lack of clear policy for crop patterns is a challenge facing optimum utilization of this water source. In general crop patterns have to be established by the MoA based on different factors. Although the MoA took some attempts in guiding farmers to establish crop patterns, it is believed that those attempts were not comprehensive enough. Marketing, for instance, was absent in the adopted approach, which resulted in losing the trust of the farmers. The farmers did not agree with the proposed cropping patterns. Secondly, existing extremely low irrigation water tariffs did not support optimum utilization of water quantity. Current tariff is 0.014 US$ per cubic meter, which presents a real barrier against water conservation or optimum utilization.

Several solutions could have been employed in order to enhance the implementation of wastewater polices and reuse standards. Among them, the following can be of priority:

1. Coordination plans should be established between different stakeholders. For instance, monitoring programs can be the responsibility of one body and results can then be shared between different regulatory bodies. An alternative would be to utilize the capacities of each governmental body for partial monitoring while data from different monitoring bodies can still be compiled and shared between them in order to maximize utilization of limited financial resources. Coordinated decisions can then be made.
2. Public awareness campaigns should be intensified in order to appreciate reclaimed water value. Special training programs should be directed to farmers aiming at introducing best agricultural practices that can be implemented to optimize water use but also quality of product and marketing of harvests.

5.2 *The Impact*

When considering the direct use of reclaimed water most farmers apply furrow or border irrigation. This is basically due to the fact that irrigation is limited so far to fodder crops, olive trees or other fruit trees. Only farms utilizing effluent of Wadi Musa WWTP (see Table 2) apply drip irrigation systems as part of a project funded by the USAID to serve Petra City and surrounding villages. In fact, water tariffs which do not exceed 10 fils/m^3 (0.014 \$US/m^3) hinder water conservation in such farms, and consequently more efficient irrigation water systems are not encouraged. Other discouraging factor can be related to farmers demand to maximize their financial incomes.

In the Jordan Valley, farmers mainly follow indirect use via means such as drip irrigation systems with plastic covers in order to avoid excessive evaporation as shown in Fig. 1. This practice positively influences the microbiological safety of

Table 2 Use of treated effluent at or near wastewater treatment plants

WWTP	Effluent quantity (MCM/yr)[a]	Amount of reused effluent (directly and indirectly) (MCM/yr)[a]	Irrigated area at or near WWTPs (Dunum)[a]	Type of irrigated crops[a]	No. of agreements with farmers[b]	% of direct reuse of treated wastewater[b]	Destination of excess effluent[a]	% of direct and indirect reuse of treated effluent[a]
As Samra	87	87	3990	Fodder and olive trees	34	15	King Talal Dam	100
Al-Fuheis	0.8	0.8	30	Fodder	1	4	Wadi Shu'aib Dam	100
Al-Ramtha	1.4	1.4	1302	Fodder	22	100	–	100
Madaba	1.8	1.8	1213	Fodder and olive trees	27	100	–	100
Al-Baq'a	4.1	4.1	437	Nurseries and a polo field	15	13.6	King Talal Dam	100
Kufranja	0.9	0.9	812	Forest trees	10	100	–	100
Al-Karak	0.7	0.7	609	Fodder and forest trees	8	100	–	100
Al-Mafraq	0.6	0.6	660	Fodder	18	100	–	100
Al-Salt	2.2	2.2	100	Olive and fruit trees	5	4.4	Wadi Shu'aib Dam	100
Ma'an	0.8	0.4	357	Fodder	9	47	Stream	47
Al-Ekeider	1.0	0.961	1069	Olive and fruit trees	17	100	–	100
Al-Sharee'a[c]	0.1	0.1	181	Olive and fruit trees	16	100	–	100

(continued)

Table 2 (continued)

WWTP	Effluent quantity (MCM/yr)[a]	Amount of reused effluent (directly and indirectly) (MCM/yr)[a]	Irrigated area at or near WWTPs (Dunum)[a]	Type of irrigated crops[a]	No. of agreements with farmers[b]	% of direct reuse of treated wastewater[b]	Destination of excess effluent[a]	% of direct and indirect reuse of treated effluent[a]
Wadi Al-Seer	1.2	1.2	62	Olive trees	1	4.3	Al-Kafrain Dam	100
Wadi Hassan	0.4	0.4	721	Olive and fruit trees	1	100	–	100
Wadi Mousa	0.9	0.9	1069	Fodder and olive trees	38	100	–	100
Abu Nuseir	0.9	0.18	75	Ornamentals	1	20	Bereen Stream	22
Al-Aqaba/ natural plant	2.0	2.0	1580	Palm trees, windbreaks, and green areas	4	100	–	100
Al-Aqaba/ mechanical plant	2.6	2.6	–	Green areas[d]	1	100	–	100
Al-Tafileh	0.5	–	–	–	None	0	Ghor Fifa	0
Al-Lajoon	0.3	–	–	–	None	0	Al-Lajoon Stream	0
Wadi Al-Arab	3.7	–	–	–	None	0	Jordan River	0
Al-Talibeyeh	0.1	0.1	–	Forest trees and ornamentals	None	100	–	100

(continued)

Table 2 (continued)

WWTP	Effluent quantity (MCM/yr)[a]	Amount of reused effluent (directly and indirectly) (MCM/yr)[a]	Irrigated area at or near WWTPs (Dunum)[a]	Type of irrigated crops[a]	No. of agreements with farmers[b]	% of direct reuse of treated wastewater[b]	Destination of excess effluent[a]	% of direct and indirect reuse of treated effluent[a]
Tal Al-Mantah	0.1	0	–	Fodder	None	0	–	0
Al-Mi'rad	0.3	0.3	–	–	1	0	King Talal Dam	100
Central Irbid	3.0	–	–	–	None	0	Jordan River	0
Jarash	1.2	1.2	–	–	None	0	King Talal Dam	100
Wadi Shalala[b]	0.8[b]	–	–	–	None[b]	0[b]	Jordan River[b]	0[b]

[a]*Source* WAJ (2012)
[b]*Source* WAJ (2013)
[c]Desalination treatment plant
[d]The effluent is also used for industrial purposes

Fig. 1 General irrigation system and scheme in the Jordan Valley with indirect use of treated wastewater through drip irrigation

the crops as well since there is no contact between irrigation water and planted crops.

As described earlier, the indirect use of treated wastewater for irrigation is taking place mostly in the middle and southern Jordan Valley (ACWUA 2011). The indirect use is for unrestricted irrigation (Ammary 2007). Nevertheless, supply of fresh water to Jordan Valley Authority (JVA) is increasingly declining because of the reduced stream flow in the Yarmouk River and side wadis and reduced rainfall in the Jordan River watershed (ISSP 2012). The types of crops which are indirectly irrigated with treated wastewater in the middle and southern Jordan Valley include grapes, vegetables, citrus, bananas, and certain types of stone fruits (Ammary 2007). According to JVA and Ministry of Agriculture MoA (2010), 212,525 Donum (1 Donum = 0.1 ha) of land in the Middle and Southern Jordan Valley were indirectly irrigated with reclaimed water during the year 2010. There are some violations to the reuse standards JS 893/2006 which occur alongside streams located downstream WWTPs where farmers use the reclaimed water discharged in these streams for unrestricted crop irrigation prior to the process of blending reclaimed water with freshwater in the dam (Ghneim 2010).

On the other hand, around 23.82% of the treated wastewater produced at treatment plants was directly used for irrigation in 2013 (WAJ 2013). Table 2 shows details pertaining to the direct use of treated wastewater for irrigation at or near each WWTP such as the type of irrigated crops and the planted area. As seen in Table 2, the overall amount of treated effluent produced by WWTPs in the year 2012 was 118 MCM and the overall area irrigated with treated effluent at the

WWTPs was 14266 Donum that same year (WAJ 2012), which represents around 6% of the total land irrigated either directly or indirectly with treated wastewater. Table 2 also shows the number of agreements between farmers and the water authority of Jordan. These agreements specify the conditions according to which treated wastewater is directly used for irrigation near WWTPs.

Drip irrigation systems are more efficient but expensive compared with furrow or border irrigation systems. Additionally, drip irrigation systems have to be replaced on regular basis averaged 5 years. When farmers are only allowed for restricted irrigation, which does not create as much income as products of unrestricted irrigation (Majdalawi 2003), they are discouraged to invest in more efficient water systems. Accordingly, shifting to higher value crops that have better financial returns to farmers is a win-win situation that will result in applying more efficient irrigation water systems and probably better acceptance for higher water tariffs. This can be possible, if risks associated with reclaimed water use are carefully managed as proposed by WHO (2006) guidelines and adapted recently by JSMO through its publication JS 17:66/2014.

6 Concluding Remarks

Jordan is one of the few countries in the West Asia that manages its scarce water resources quite well. The country has 97% water network coverage, which is one of the highest at regional level. In addition, around 63% of the population is served with sewerage network. All collected wastewater is treated to at least a secondary level, while more than 70% of this water is treated up to a tertiary level. More than 90% of treated wastewater is used mainly for irrigation purposes. Recently developed policies clearly show the governmental willingness to increase the sanitation services and expand treated wastewater agricultural reuse. Both centralized and decentralized wastewater treatment are considered with a focus on centralized approaches. A separate policy was published on 2016 for decentralized wastewater management in which 28% of the population is targeted. The government is currently developing action plans for policy implementation.

Controlled treated wastewater use for agricultural production had been practiced in Jordan since decades. The original motivation for such practice was the severe shortage in fresh water resources, which affected the agricultural sector. In fact, the recently introduced policies clearly states that fresh water used for irrigation would be gradually substituted with treated wastewater whenever feasible. Wastewater treatment in Jordan received an early attention and as a result, it is highly regulated and controlled now. Notwithstanding the required coordination, many authorities have the mandate of controlling effluent quality and taking all necessary actions to ensure both safety of products and public health. High and unnecessary restrictions on reclaimed wastewater use in agriculture was enforced; and direct effluent use is so far only allowed for fodder crops and for trees. Moreover, effluents of wastewater treatment plants can under no circumstances be used for unrestricted irrigation

(irrigation of crops that can be eaten raw). Unrestricted irrigation is only allowed after mixing treated effluent with other fresh water resources. Although, the practice of wastewater irrigation is well developed in the country, there is still much to be done in order to maximize the benefits. The sector is highly subsidized and therefore, the tariffs have to be reconsidered. This is also applicable for all other fresh water resources; hence, the recent water strategy (2016–2025) considered tariffs calculations based on water allocation models. Moreover, maximizing benefits may be achieved by moving towards cash crops. This requires better implementation of standards by Water Authority of Jordan through removing the unnecessary restrictions put on reclaimed wastewater use for irrigation, and perhaps reconsideration of the enforced standards. In addition, adoption of the standards JS 1766/2014, which was developed based on WHO (2006) guidelines is highly recommended through development of sanitation safety plans at catchment area level.

References

Abuzeid, K. (2014). An Arab perspective on the applicability of the water convention in the Arab region: key aspects and opportunities for the Arab countries. In *Workshop on Legal Frameworks for Cooperation On Transboundary Water. Tunis, 11–12 June, 2014.*

Abuzeid, K., & Elrawady, M. (2014). 2nd Arab State of the Water Report. Center for Environment and Development for the Arab Region and Europe and Arab Water Council.

AFED. (2014). Water efficiency handbook: Identifying opportunities to increase water use efficiency in industry, buildings, and agriculture in the Arab countries.

Almazroui, M. (2012). Dynamical downscaling of rainfall temperature over the Arabian Peninsula using RegCM4. *Climate Research, 52*, 49–62. http://www.int-res.com/articles/cr_oa/co52po49.pdf.

Ammary, B. (2007). Wastewater reuse in Jordan: Present status and future plans. *Desalination, 211,* 164–176.

Arab Countries Water Utilities Association (ACWUA). (2011). *Safe use of treated wastewater in agriculture: Jordan case study*. Amman, Jordan: Nayef Seder (JVA) and Sameer Abdel-Jabbar (GIZ).

GEO-6. (2016). *Global environment outlook. Regional assessment for West Asia*. United Nations Environment Program (UNEP). ISBN 978-92-807-3548-2.

Ghneim, A. (2010). *Wastewater reuse and management in the Middle East and North Africa: A case study of Jordan*. Ph.D. Dissertation, Technical University of Berlin, Germany.

Glass, N. (2010). The water crisis in Yemen: Causes, consequences and solutions. *Global Majority E-Journal, 1*(1), 17–30.

Haddadin, M., & Shteiwi, M. (2006). Linkages with social and cultural issues. In: M. J. Haddadin (Eds.), *Water resources in Jordan: evolving policies for development, the environment, and conflict resolution. Issues in water resource policy*, (pp. 2011–235). Washington, DC: Resources for the Future.

ISSP. (2012). *Water valuation study program*. USAID/Jordan: Institutional Support and Strengthening Program (ISSP)

ISSP. (2014). National Strategic Wastewater Master Plan Final Report. http://pdf.usaid.gov/pdf_docs/PA00JRPX.pdf. Accessed 30 September 2017.

Jordan Valley Authority (JVA) and Ministry of Agriculture (MoA). (2010). Annual Report. JVA and MoA, Amman, Jordan

Kone, D. (2010). Making urban excreta and wastewater management contribute to cities' economic development: A paradigm shift. *Water Policy, 12*(4), 602–610.

Majdalawi, M. (2003). *Socio-economic and environmental impacts of the reuse of water in agriculture in Jordan. Farming systems and resources economics in the tropics No 51.* Dissertation. Hohenheim University, Stuttgart, Germany.

McCornick, P. G., Hijazi, A., & Sheikh, B. (2004). From wastewater reuse to water reclamation: progression of water reuse standards in Jordan. In: C. Scott, N. I. Faruqui, & L. Raschid. (Eds.), *Wastewater use in irrigated agriculture: confronting the livelihood and environmental realities*, Wallingford: CABI/IWMI/IDRC.

MEDAWARE. (2005). *Development of tools and guidelines for the promotion of the sustainable urban wastewater treatment and reuse in the agricultural production in the mediterranean countries.* Project Acronym (MEDAWARE), Task 5: Technical Guidelines on Wastewater Utilisation.

Meslemani, Y. (2008). *Climate change impacts and adaptation in the eastern Mediterranean/ Syria: Draft UNFCCC initial national communication for Syria.* Demascus, Syria: Ministry of State for Arab Affiars.

Ministry of Water and Irrigation. (2016). *National water strategy of Jordan, 2016–2025.* http://www.mwi.gov.jo/sites/en-us/Hot%20Issues/Strategic%20Documents%20of%20%20The%20Water%20Sector/National%20Water%20Strategy(%202016-2025)-25.2.2016.pdf. Accessed on September 30, 2017.

MWI (2009). *Water for life: Jordan's water strategy 2008–2022.* http://www.mwi.gov.jo/sites/en-us/Documents/Jordan_Water_Strategy_English.pdf.

Nazzal, Y. K., Mansour, M., AL Najjar, M., & McCornick, P. (2000). *Wastewater reuse laws and standards in the Kingdom of Jordan.* Amman, Jordan: The Ministry of Water and Irrigation.

Olmstead, S. M. (2014). Climate change adaptation and water resource management: A review of the literature. *Energy Economics, 46,* 500–509.

Rachel, B., Jeanne, L., & Jamie, B. (2013). Sanitation: A global estimate of sewerage connections without treatment and the resulting impact on MDG progress. *Environmental Science and Technology, 47*(4), 1994–2000.

Reymond, P., Renggli, S., Luthi, C. (2016). Towards sustainable sanitation in an urbanizing world. In: M. Ergen (Ed.), *Chapter 5 in the bool sustainable urbanization.* Intech Printing. ISBN 978-953-51-2653-9, Print ISBN 978-953-51-2652-2.

Rowling, M. (2014). *Iraq's environment water supply in sever decline.* Washington: Thomson Reuters Foundation News.

Sipkin, S. (2012). *Water conflict in Yemen. ICE case studies (235).* http://www.1.american.edu/ted/ice/yemen-water.htm.

Ulimat, A. (2012). Wastewater production, treatment, and use in Jordan. In: *Second Regional Workshop: Safe Use of Wastewater in Agriculture, New Delhi, India*, 16–18 May 2012.

UNEP. (2016). *Global environment outlook: Regional assessment for West Asia.* United Nations Environment Program (UNEP). IBSN 978-92-807-3548-2.

Van Vliet, B., Spaargaren, G., & Oosterveer, P. (Eds.). (2010). *Social perspectives on the sanitation challenge.* Dordrecht: Springer. ISBN 978-90-481-3721-3.

Water Authority of Jordan (WAJ). (2012). *Quantities of Treated Wastewater Exiting WWTPs and Used Directly and Indirectly for Irrigation.* Technical Report. Amman, Jordan: Water Reuse and Environment Unit, WAJ

Water Authority of Jordan (WAJ). (2013). *Agreements with farmers for purposes of reusing treated wastewater in irrigation.* Technical Report. Amman, Jordan: Water Reuse and Environment Unit, WAJ

WHO (2006). *Guidelines for the safe use of wastewater, excreta and grey water* vol 2. wastewater use in agriculture. Published by the World Health Organization. ISBN: 92 4 154683 2.

Development of Sanitation Safety Plans to Implement World Health Organization Guidelines: Jordanian Experience

Maha Halalsheh, Ghada Kassab, Khaldoun Shatanawi and Munjed Al-Shareef

Abstract Although the World Health Organization (WHO) guidelines for use of wastewater, greywater and excreta in agriculture were available for quite some time, such guidelines were not adopted by most countries due to difficulties in implementation. A critical step for rendering such guidelines applicable is the development of a detailed implementation plan, which had been recently recognized as Sanitation Safety Planning (SSP). The plan shall ultimately define roles and responsibilities of different authorities and other role players in the whole process. A manual that presents a step-by-step guidance for the development of the SSP was recently published by WHO. A proper framework is often necessary for a country to develop the detailed Sanitation Safety Plans (SSPs). The framework should formulate roles and responsibilities associated with the implementation process and introduce them to different authorities for feedback and approval before investing in a detailed SSP. This chapter first introduces SSP in brief and its framework and then further presents examples from Jordan in which the WHO 2006 guidelines were validated and a framework for SSP was developed in consultation with different authorities. Jordan has relatively advanced experience in wastewater use in irrigation stemming from the high demand for managing its extremely limited water resources. Accordingly, the country specified guidelines JS1766/2014 that adapts the WHO 2006 wastewater management approaches to aim for the full utilization of resources. The new guidelines allow the use of wastewater for unrestricted irrigation. Restrictions were shifted to agricultural practices and other downstream practices that are believed to guarantee a produce complying with the current

M. Halalsheh (✉)
Water, Energy and Environment Center, The University of Jordan, Amman, Jordan
e-mail: halalshe@ju.edu.jo; nameer_maha@hotmail.com

M. Halalsheh
GIZ, Amman, Jordan

G. Kassab · K. Shatanawi
Faculty of Engineering, University of Jordan, Amman, Jordan

M. Al-Shareef
Faculty of Engineering, German Jordanian University, Madaba, Jordan

H. Hettiarachchi and R. Ardakanian (eds.), *Safe Use of Wastewater in Agriculture*,
https://doi.org/10.1007/978-3-319-74268-7_6

standards. The framework developed in this context supports the current standards and laid the foundation for the development of detailed SSPs.

Keywords Wastewater · Greywater · Excreta · Agriculture · Sanitation safety plans (SSPs) · Sanitation safety planning (SSP) · Guidelines · Implementation framework

1 World Health Organization Guidelines for Safe Use of Wastewater

Safe wastewater use in agriculture had been so far "guided" by the World Health Organization (WHO) guidelines (WHO 1989), which stipulated quality parameter limits for effluents of wastewater treatment plants (WWTPs). Although the guidelines tackled health risks related to pathogens existing in wastewater, end of pipe technologies were always deemed as the basis for safe water use. Explicitly, maximum permissible values were set to determine the quality of treated water that can be used for agricultural irrigation. There are two main drawbacks in such approach. Firstly, wastewater that receives treatment is estimated to be at only 10% (Murtaza et al. 2010) on a global scale. This means that about 90% of wastewater is either reused indirectly, after being discharged to waterbodies, or directly in irrigated agriculture. Such practice is neither controlled nor guided and apparently not covered by the WHO guidelines (1989). Secondly, there is evidence for effluent contamination or regrowth of pathogens downstream of WWTPs, or in effluents stored before being reused in agricultural production. Therefore, setting quality parameters are not merely sufficient to guarantee safe water reuse downstream of the treatment plant.

Above issues demanded, a dramatic shift in the ways how wastewater use in agriculture should better controlled is necessary. WHO guidelines published in 2006 was a result of such demands. A clear shift in wastewater management approach was observed in the 2006 guidelines (WHO 2006), including the need to involve different stakeholders in determining the risks and risk mitigation strategies. The guidelines addressed WWTP effluent quality in conjunction with agricultural practices aiming at safe reuse of different wastewater qualities. Figure 1 shows how WHO shifted its borderline from the downstream of treatment plant towards agricultural fields and further along the rest of the food chain. Farming practices are of utmost importance in this integrated approach; in which minimally treated wastewater was not excluded from being safely used in agriculture. It should be noted that other farming practices that may have an impact on produce is to be furthermore considered. For instance, pesticides application may result in non-communicable diseases, as it is the case for organochlorine pesticides, which are known to be carcinogenic. Such group of pesticides was shown to accumulate in soil and easily enter food chain (Batarseh and Tarawneh 2013). Consequently, it is indispensable to consider best farming practices in combination with other practices related to treated wastewater use as presented by the WHO guidelines (2006). Another relevant example is related to produce contamination by pathogens present in unprocessed manure that is used as fertilizer. In fact, agricultural irrigation using fresh water does not mean a produce complying with imposed standards since water quality is not the only determinant of the produce quality. Accordingly, it is believed that agricultural use of wastewater should be considered in a comprehensive perspective into which water quality is one element. Other input variables are as important as water such as fertilizers quality and pesticides application.

Notwithstanding integrity of the full system proposed by WHO guidelines (2006), the absence of detailed management plan limits its applicability. Obviously, management plans are expected to vary from country to country, as well as within the same country, depending on different variables. Emphasis should be given to the role of coordination between different stakeholders when developing applicable management plan, but also at implementation stage. Plans can be established for the whole sanitation chain, or can be progressively developed according to existing conditions. Moreover, management plans can be designed to deal with acute conditions when raw sewage is used for agricultural production (e.g. focus on risk management of microbial hazards); while more comprehensive plan can be developed when wastewater is well treated, in which good agricultural practices may deal with additional chemical hazards. In any case, the main two pillars of the approach are: firstly, ensuring the public health and secondly ensuring the produce safety. To a lesser extent, impact of implementing such approaches on environment may be considered.

The management plans proposed to accomplish above and implement WHO guidelines (2006) are known as sanitation safety plans (SSPs). The SSPs prioritize risks and utilize limited resources to target highest risk allowing for progressive improvements. A manual has been developed recently to provide step-by-step guidance to assist the implementation of the WHO guidelines (2006) for the Safe

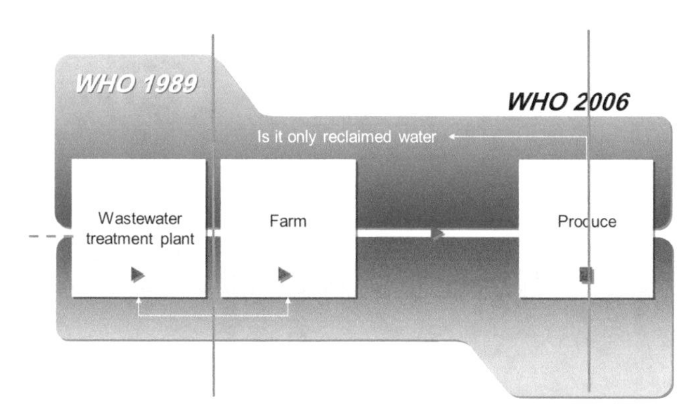

Fig. 1 Paradigm shift in sanitation approach from end of pipe technology (WHO 1989) to integrated management approach (WHO 2006)

Use of Wastewater, Excreta and Greywater in agriculture (WHO 2015a, b). Development of a framework that can enhance the understanding of the system and facilitate precise development of detailed SSPs, is considered a step prior to the Sanitation Safety Planning (SSP). The framework should provide the institutional conceptual structure needed for the SSP and serve as informative tool for relevant authorities.

In this context, this chapter aims at describing the steps embedded in sanitation safety planning and presents an example from Jordan in which WHO guidelines (2006) were tested under local conditions and framework for SSP was developed. The Jordanian experience presented in here confirms the necessity of developing and adopting SSPs even when wastewater quality is controlled by treatment.

2 Sanitation Safety Planning (SSP)

Development of SSPs is modeled after the Stockholm framework for preventive risk assessment and management. It follows almost the same approach used in the development of Water Safety Plan (WSP) (Davison et al. 2005). Similar to WSP, SSPs also comprises three main components: system analysis and design, operational monitoring and management plans as shown in Fig. 2. Each of these components as well as the supporting programs needed, are briefly introduced in the following subsections.

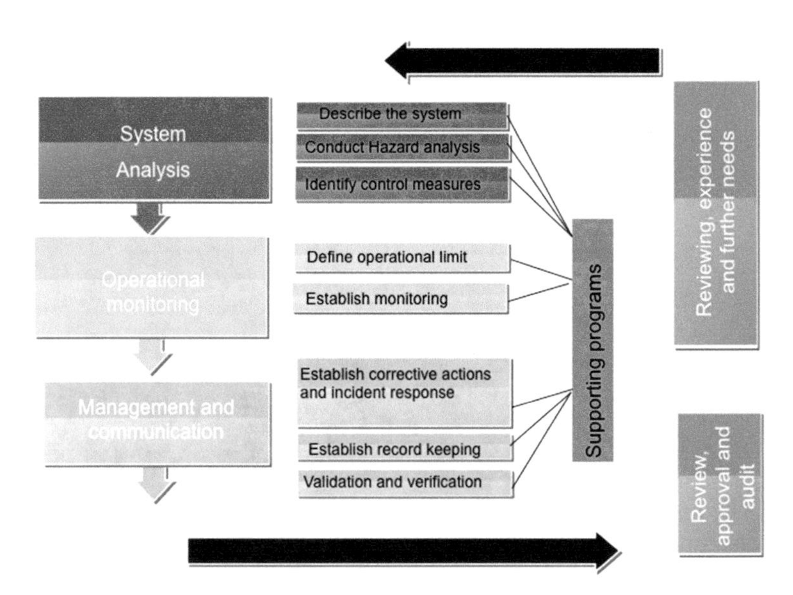

Fig. 2 Components of SSPs; adapted from Davison et al. (2005)

2.1 System Analysis

System analysis consists of the three consecutive steps:

(1) **System description**, which covers the whole chain (from the toilet to the farm and then to the table) and can be best represented by flow chart that carefully delineates the system;
(2) **Hazard analysis** in which identification of all potential hazards (biological, chemical, physical, and radiological agents that have the potential to cause harm), their sources, possible hazardous events and an assessment of risk presented by each (Davison et al. 2005); and
(3) **Control measures**, which are steps needed along the chain in order to ensure that health based targets are met. They are actions or activities that have to be applied to minimize hazards.

For instance, at the farm level, applying drip irrigation system would present a barrier to microbial hazard transfer. Alternatively, other barriers (control measures) can be applied as shown in Fig. 3 and will be further elaborated later. Control measures and frequency of monitoring should reflect likelihood and consequences of the loss of control. In any system, there may be many hazards and potentially a large number of control measures. It is therefore important to rank the hazards in order to establish priorities (Davison et al. 2005).

2.2 Operational Monitoring

It is important to define the operational limits that lead to the safe practices. Operational limits should not necessarily mean concentration of hazard, but rather a gauge of control measure performance that can explain the objective of monitoring.

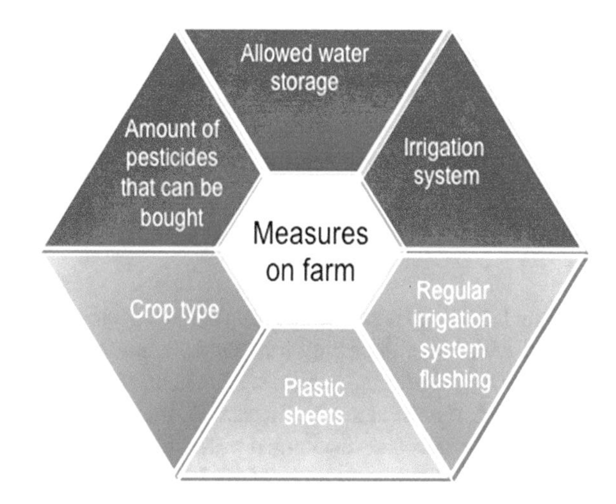

Fig. 3 Examples of control measures (barriers) that can be implemented at farm level

Performance monitoring relies on establishing 'what', 'how', 'when' and 'who' principles (Davison et al. 2005). The objective of monitoring is to monitor control measures in timely manner to prevent wastewater from being used unsafely in agriculture. A monitoring program should be established and records of all monitoring shall be maintained.

2.3 Management and Communication

When monitoring indicates a deviation from the established operational limit, there is a need for corrective action in order to restore operation and ensure safety of wastewater use in agriculture. Clear descriptions of actions to be taken during such situation should be provided. Moreover, appropriate documentation and reporting has to be established.

2.4 Supporting Programs

Supporting programs comprise all activities that ensure process control such as standard operating procedures, hygienic practices, and raising awareness among the communities. Accordingly, supporting programs are not directly part of SSP; however, they are extremely important in maintaining the operating environment and ensuring proper control.

Since aforementioned description is still theoretical, a better clarification can be achieved through a practical example. The rest of the chapter is aimed at providing such example in which Jordan may serve as a model where a well-defined policy exists and encourages agricultural wastewater reuse. Jordan standard and metrology organization (JSMO) has recently issued standards JS 1766-2014 for wastewater reuse in agriculture based on WHO guidelines (2006). Issued standards indeed need detailed implementation plan, which still does not exist. As a first step, framework for SSPs was developed and will be further used for advancing a detailed implementation plan.

3 Development of the Framework for SSP in Jordan

Jordan has an area of around 89,000 km^2. More than 94% of the population is served with water distribution network, while sewage networks cover around 63% (MWI 2016). Collected wastewater is being treated and almost all effluent is being reused in agricultural sector. In year 2014, 128 million cubic meters (MCM) were treated in Jordan and discharged either into watercourses or used directly for irrigation or other intended uses (MoE 2016). This volume is expected to increase to

235 MCM by year 2025 (MWI 2016) and will contribute to 16% of the total water budget. About 70% of the collected wastewater is treated in As-Samra wastewater treatment plant, which is equipped with tertiary treatment for nitrogen removal. The plant serves Amman and Zarqa cities where half of the population is living. The remaining 30% of collected wastewater is treated in 27 treatment plants where at least secondary treatment is applied with BOD removal efficiencies nearing 95% in most cases.

Treated wastewater is being used for irrigation directly or indirectly (i.e. after mixing with surface water). Indirect use mostly takes place in middle and southern Jordan Valley (ACWUA 2011; Carr et al. 2011). Indirect use is practiced for unrestricted irrigation (Ammary 2007). According to the Jordan Valley Authority (JVA) and the Ministry of Agriculture (MoA), 21,253 ha of land was indirectly irrigated with reclaimed water in 2010 (JVA and MoA 2010). On the other hand, about 24% of treated wastewater was being directly used for irrigation in 2013 (WAJ 2013). Most farmers apply furrow or border irrigation in direct reclaimed water use. This is due to the fact that irrigation is limited to fodder crops, olive trees or other fruit trees. In fact, water tariffs which do not exceed 10 fils/m^3 (0.014$US) hinder water conservation in such farms, and consequently discourage the use of more efficient irrigation water systems. Other discouraging factor can be related to farmers demand to maximize their financial gain. More efficient irrigation systems, such as drip irrigation (a control measure suggested in SSPs), are indeed more expensive. Additionally, drip irrigation systems have to be replaced on a regular basis of approximately 5 years. Since farmers are only allowed to practice restricted irrigation, which does not create as much income as products of unrestricted irrigation (Majdalawi 2003), they are discouraged to invest in more efficient water systems. Accordingly, shifting to higher value crops that have better financial returns to farmers is a win-win situation that will result in applying more efficient irrigation water systems and probably better acceptance for higher water tariffs. This may be possible, if risks associated with reclaimed water use are carefully managed as proposed by WHO (2006) guidelines.

In 2013, a consortium consisting of the University of Jordan and German Jordanian University was assigned by the WHO to conduct a study aiming at development of framework for SSPs for Jordan. The objectives of the study were to;

- Validate WHO guidelines (2006) within Jordanian context through experiments; and
- Consequently use above results to formulate the required framework.

The study was neither meant to work on setting health-based targets, quantitative microbial risk analysis and other risk assessment approaches, nor looked at synthesis of risk assessment. It rather selected the most conservative target, which is reduction of 6logs of *E. coli* and aimed at the best possible outcome.

The study conducted baseline analysis showing existing situation of wastewater management in Jordan and defined roles and responsibilities of each stakeholder. Hazards associated with both reclaimed water reuse and with other existing agricultural practices were also identified and prioritized. The study did not go into identifying disease pathways and affected groups of people. Such identification is believed to be part of detailed SSPs. Study borders started at WWTP effluent and focused on practices at farm level as shown in Fig. 4. It should be noted that the whole sanitation chain was not the target of this study. Especially, the elements starting from transportation and ending at consumer hands are not limited to the produce irrigated with treated wastewater. Additionally, neither industrial wastewater nor sludge produced by WWTPs was considered in this study, although both can be utilized in agriculture.

3.1 *Designing Experiment for WHO (2006) Guidelines Validation*

In order to validate WHO (2006) guidelines for the Jordanian context, two experiments were conducted. Their objectives and descriptions are provided in the following subsections.

3.1.1 First Experiment

The first experiment was designed to examine the main source of raw vegetables contamination. This was conducted in two open field farms;

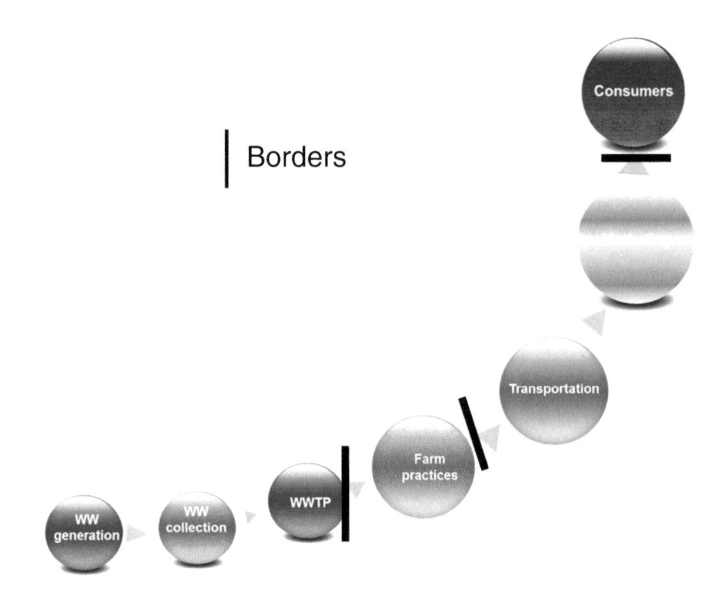

Fig. 4 Borders of implemented study

(1) The first farm was located on the banks of Zarqa River used as the source of irrigation water in this area. Treated wastewater discharged from Kherbit As-Samra wastewater treatment plant contributes to most of river annual yield. The first farm was cultivated with Zucchini, cabbage and bell pepper. For each type of crops, two 50 m rows were cultivated. The distance between rows was 1.2 m and plant distance within each row was 0.40 m. The soil characteristics measure before cultivation, are shown in Table 1. The three types of crops received the same type of irrigation water. Cultivation followed farming practices adopted in the area, except for manure application, as no manure had been applied to the plots during the whole experimental season.
The first farm was irrigated with water discharged directly from Zarqa River with quality shown in Table 2. Above ground drip irrigation system was applied. Drip lines were placed on the middle of each row, with 40 cm distance between drip emitters.
(2) The second farm that was irrigated with ground water was located at Al-Mafraq governorate (32 20 04.71 N; 36 18 28.19 E) to the north east of Jordan and was using water with drinking water quality for irrigation. The second farm was cultivated with Zucchini, cabbage and bell pepper. For each crop an area of 40,000 m^2 was cultivated. Each 20,000 m^2 was served with a main that supply irrigation water to 100 rows on each side. Distance between rows was 1.2 m and distance between plants within each row was 0.4 m.

Table 1 Soil characteristics of Zarqa River pilot farm

Parameter	Unit	Sample I		Sample II		Sample III	
		Depth 0–20 cm	Depth 20–40 cm	Depth 0–20 cm	Depth 20–40 cm	Depth 0–20 cm	Depth 20–40 cm
Soil texture		Clay loam					
Sand	%	37.30	41.40	37.40	38.10	31.70	33.60
Silt		28.10	26.00	32.40	29.90	34.10	32.10
Clay		34.60	32.60	30.20	32.00	34.10	34.30
organic carbon		3.00	2.40	2.00	2.40	3.00	2.80
TN		0.17	0.14	0.16	0.16	0.16	0.16
pH (1:2)	SU	8.13	8.09	8.24	8.24	8.13	7.72
Ec at 25C (1:2)	μs/cm	431.00	488.00	285.00	366.00	1004.00	1923.00
B	mg/kg (dry)	4.08	3.75	<0.24	4.05	5.29	5.43
Ca (exchangeable and extractable)		5420.00	5520.00	5380.00	5202.00	5420.00	5310.00
Mg (exchangeable and extractable)		1520.00	1440.00	1592.00	1545.00	1763.00	1994.00
Na (exchangeable and extractable)		122.00	297.00	79.10	135.00	461.00	716.00
TCC	MPN/gm	<0.3	<0.3	<0.3	<0.3	<0.3	<0.3

Table 2 Irrigation water characteristic/Zarqa River pilot farm

Parameter	Unit		JS (893/2006)[a]
pH		8.2 (6)[b]	6–9
Total suspended solids (TSS)	mg/l	20.2 (6)	50
Total dissolved solids (TDS)	mg/l	1157.0 (6)	1500
COD	mg/l	57.3 (6)	100
BOD	mg/l	6.8 (6)	30
Sodium	mg/l	208.3 (6)	230[c]
Calcium	mg/l	53.5 (6)	230[c]
Magnesium	mg/l	25.8 (6)	100[c]
Potassium	mg/l	34.5 (6)	
Total N	mg/l	15.9 (6)	45
Ammonium as NH_4	mg/l	0.3 (6)	–
Nitrate as NO_3	mg/l	47.6 (6)	30
Carbonate	mg/l	3.2 (6)	
Total alkalinity	mg/l	204.0 (6)	
Boron	mg/l	8.2 (6)	1.0[c]
E. coli	MPN/100 ml	2353 (4) >160,000 (1) >1600 (1)	100
Total coliform	MPN/100 ml	1.4E + 04 (4) >160,000 (1) >1600(1)	

[a]For irrigation of vegetables eaten cooked
[b]Values between brackets are number of samples
[c]Guidelines values

Al-Mafraq farm was irrigated with groundwater possessing quality shown in Table 3, and also by means of above ground drip irrigation system. Drip lines were placed in the middle of each plantation row, with 40 cm distance between drip emitters.

3.1.2 Second Experiment

This was to examine the health protection measures established by WHO guidelines (2006) for the Safe Use of Wastewater, Grey water and Excreta in agriculture. The second experiment was conducted in a greenhouse, located at Abu-Nussier Wastewater Treatment Plant. The greenhouse had no heating, and had passive ventilation with lateral side panels. The cropping area was 200 m^2. Selected crops were high growing tomato and lettuce. The greenhouse was organized into 12 plots, measuring (3 m by 4 m). Six plots were cultivated with high growing tomato and irrigated with surface drip irrigation while the other six plots were cultivated with lettuce and irrigated with subsurface irrigation. A two to three week old transplants

Table 3 Irrigation water characteristic/Mafraq farm

Parameter	Unit	
pH		7.9
Total suspended solids (TSS)	mg/l	<5
Total dissolved solids (TDS)	mg/l	478
COD	mg/l	15
BOD	mg/l	<3
Sodium	mg/l	111
Calcium	mg/l	49
Magnesium	mg/l	45
Potassium	mg/l	<0.4
Total N	mg/l	5.2
Parameter	Unit	
Ammonium as NH_4	mg/l	1.1
Nitrate as NO_3	mg/l	14.54
Bicarbonate	mg/l	146
Carbonate	mg/l	<2.5
Total alkalinity	mg/l	120
E. coli	MPN/100 ml	
Total coliform	MPN/100 ml	<1.8

were cultivated with plant density of 2.5 plant/m^2. Soil characteristics are presented in Table 4.

Three types of water quality were used for irrigation as shown in Table 5: (i) secondary treated effluent (SE), (ii) disinfected secondary effluent (DSE) and (iii) fresh water. Each type of irrigation water was used for irrigation of two plots of each cultivated crop, i.e. each type of irrigation water was used for irrigation of two tomato plots and two lettuce plots. Surface drip irrigation was used for cultivation of high growing tomato. Each tomato plot comprised three rows of plants and accordingly three irrigation lines with 1.2 m spacing between lines and 0.4 m spacing between drip emitters within drip lines, giving an emitter density of 2.5 emitters/m^2. The drip emitters had discharge rate of 4 l/h. Subsurface drip irrigation was used for cultivation of lettuce. Each plot comprised three rows of plants. Subsurface drip lines were placed in the center of each plant's row and buried at 10 cm depth. The distance between drip emitters were 40 cm and had a discharge rate of 1.6 l/h. Emitters were placed upward to minimize clogging problems.

3.2 *Sample Collection, Analysis and Results*

Irrigation water samples were collected over ten sampling events during the irrigation period. All samples were grab samples, which upon collection stored in a cooling box and transported to local laboratory for analysis on the day of sampling.

Table 4 Soil characteristics of Abu-Nussier WWTP pilot farm

Parameter	Unit		
		Depth 0–20 cm	Depth 20–40 cm
Soil texture		Clay	
Sand	%	25.00	27.50
Silt	%	32.50	30.00
Clay	%	42.50	42.50
Organic carbon	% dry	0.58	0.44
TN	% dry	0.07	0.06
pH (1:2)	SU	8.46	8.18
Ec at 25C (1:2)	μs/cm	374.00	593.00
B	mg/kg (dry)	22.90	23.60
Ca (exchangeable and extractable)		6242.00	6356.00
Mg (exchangeable and extractable)		1405.00	1013.00
Na (exchangeable and extractable)		321.00	380.00
TCC	MPN/gm	<3	<3

Analysis of water samples for *E. coli* was initiated on the day of collection. All parameters were analyzed according to the Standard Methods for the Examination of Water and Wastewater (APHA 2012). For the two experiments, the sampling of soil was performed before plantation, during irrigation and just before harvesting. Sampling during irrigation was coordinated with irrigation events so as soil samples were collected either on the same day of irrigation or within one to three days after an irrigation application. Soil samples were collected with an auger within a 25 cm radius of a drip emitter. The soil core was divided into an upper (0–20 cm) and a lower soil fractions (21–40 cm) which were analyzed separately. Before irrigation started, a composite sample consisting of 18 cores and 6 cores-for each soil fraction —were collected from Zarqa River pilot farm and Abu-Nussier WWTP pilot farm, respectively. During the irrigation period and at later stages of harvesting a composite sample consisting of 6 cores for each soil fraction was collected for each type of crop at the Zarqa River pilot farm. For Abu-Nussier WWTP pilot farm, a composite sample consisting of 2 cores for each soil fraction was collected from each plot during the irrigation period and at later stages of harvesting. Composite samples were stored until it was transported to RSS laboratories for analysis on the same day of sampling.

For each produce, samples were collected after two, three and occasionally four days after last irrigation event. Samples of Zucchini, bell pepper and tomato consisted of 10–12 produce. Samples of lettuce and cabbage were consisting of one produce. For Zarqa River pilot farm, samples were collected over two production stages; harvesting and packaging, except for cabbage, which was collected only over harvesting. For harvesting stage, samples were picked up by farm workers immediately into a sampling bag that was held by project staff. Moreover, and to detect potential pollution originated from farm workers, samples occasionally were

Table 5 Irrigation water characteristic/Abu Nussier WWTP pilot farm

Parameter	Unit	Fresh water	Secondary effluent	Disinfected effluent	JS (893/2006)[a]
S					
pH	Unit	7.806 (6)	7.6 (6)	6.6 (6)	6–9
Total suspended solids (TSS)	mg/l	<5 (6)	10.2 (6)	7.5 (6)	50
Total dissolved solids (TDS)	mg/l	374 (6)	727.2 (6)	821.6 (6)	1500
COD	mg/l	14 (6)	55.8 (6)	38.4 (6)	100
BOD	mg/l	<3 (6)	17.4 (6)	17.3 (6)	30
Sodium	mg/l	55.0 (6)	136.5 (6)	148.1 (6)	230[b]
Calcium	mg/l	29.2 (6)	38.1 (6)	45.7 (6)	230[b]
Magnesium	mg/l	13.7 (6)	17.594 (6)	18.0 (6)	100[b]
Potassium	mg/l	16.4 (6)	26.5 (6)	28.4 (6)	
Total N	mg/l	2.52 (6)	13.9 (6)	10.2 (6)	45
Ammonium as NH_4	mg/l	0.3 (6)	3.6 (6)	7.7 (6)	–
Nitrate as NO_3	mg/l	6.9 (6)	5.7 (6)	1.6 (6)	30
Carbonate	mg/l	<2.5 (6)	<2.5 (6)	<2.5 (6)	
Total alkalinity	mg/l	100.2 (6)	207.8 (6)	99.5 (6)	
E. coli	MPN/100 ml		600 (2) >1600 (1) >16,000 (1)		100
Total coliform	MPN/100 ml	<1.8	920 (1) >1600 (2) >16,000 (1) >	<1.8	–

Values between brackets are number of samples
[a]Values presented here are defined limits within the standards. For irrigation of vegetables eaten cooked
[b]Guideline values

collected by project staff as well, wearing sterilized gloves and using sterilized equipment. As for packaging stage, project staff collected samples after it was packaged by farm workers. For reference farm, i.e. Mafraq farm, samples were collected over, harvesting, packaging and transportation stages. For Abu-Nussier WWTP pilot farm, samples were collected over harvesting stage only.

3.2.1 Results from the First Experiment

Microbial analysis for zucchini samples collected from Zarqa River pilot farm (Table 6) showed that all samples picked up after two and four days of last irrigation- whether at harvesting or packaging stages- were negative for total coliform,

Table 6 Microbiological quality of zucchini produce[a]/Zarqa River pilot farm

Parameter	After 2 days of last irrigation		After 3 days of last irrigation		After 4 days of last irrigation	
	At harvesting	After packaging	At harvesting	After packaging	At harvesting	After packaging
Total coliform (CFU/g)	0/2	0/2	1/2 2×10^3	1/2 6×10^2	0/2	0/2
E. coli (CFU/g)	0/2	0/2	1/2[a] 5×10^2	1/2 2×10^2	0/2	0/2
Salmonella (pre/abs in 25 g)	Abs/2	Abs/2	Abs/2	Abs/2	Abs/2	Abs/2

[a]Presented results are the number of samples showing total coliform or *E. coli* count above 10 CFU/g to total number of tested samples

E. coli and salmonella. Nevertheless, one of the two samples collected after three days of last irrigation, contained total coliform and *E. coli*. For cabbage, results showed (Table 7) that all collected samples were negative for total coliform and *E. coli*, except for one sample collected after irrigation-withheld period of two days. Accordingly and based on these results, we were not able to either support or refute that the adopted health measures, viz. use of surface drip irrigation, use of mulch, and allowance of pathogen die-off period of two to four days were completely effective for achieving the required pathogens reduction.

For pepper, results (Table 8) showed that *E. coli* contamination is below 10 CFU/g for all samples collected at harvesting stage, indicating that use of surface drip irrigation, use of mulch and allowing a pathogen die off period over two days is sufficient to achieve the required pathogen reduction. Nevertheless and as shown by the results, if not processed according to established protective measures, packaging process may introduce contamination. In conclusion, results have shown that because surface drip irrigation applies water at surface of soil, it is less likely for high growing crops to be contaminated as the edible parts of plant are not directly exposed to applied irrigation water. Moreover, establishing periods of no-irrigation have also contributed to the reduction of *E. coli* concentration on surface of crops to below permissible levels, i.e. 10 CFU/g.

With respect to reference farm (Al-Mafraq), results (Table 9) have shown evidence for contamination of zucchini at the three production stages, i.e. harvesting,

Table 7 Microbiological quality of cabbage produce[a]/Zarqa River farm

Parameter	After 2 days of last irrigation	After 3 days of last irrigation
	At harvesting	At harvesting
Total coliform (CFU/g)	1/4	0/4
E. coli (CFU/g)	1/4	0/4
Salmonella (pre/abs in 25 g)	–	–

[a]Presented results are the number of samples showing total coliform or *E. coli* count above 10 CFU/g to total number of tested samples

Table 8 Microbiological quality of pepper produce[a]/Zarqa River farm

Parameter	After 2 days of last irrigation		After 3 days of last irrigation		After 4 days of last irrigation	
	At harvesting	After packaging	At harvesting	After packaging	At harvesting	After packaging
Total coliform (CFU/g)	0/4	0/4	0/4	0/4	0/4	3/4
E. coli (CFU/g)	0/4	0/4	0/4	0/4	0/4	3/4
Salmonella (pre/abs in 25 g)	Abs/4	Abs/4	Abs/4	Abs/4	Abs/4	Abs/4

[a]Presented results are the number of samples showing total coliform or *E. coli* count above 10 CFU/g to total number of tested samples

Table 9 Microbiological quality of reference samples/Mafraq farm

	Zucchini			
	At harvesting		After packaging	After transportation
	Picked by farm worker	Picked by staff		
Total coliform (CFU/g)	2/4	4/8	4/4	1/2
E. coli (CFU/g)	2/4	4/8	4/4	1/2
Salmonella (pre/abs in 25 g)	Abs/4	Abs/8	Abs/4	Abs/2
Cabbage				
	At harvesting		After packaging	After transportation
Total coliform (CFU/g)	2/2	4/12	2/2	–
E. coli (CFU/g)	0/2	0/12	0/2	–
Salmonella (pre/abs in 25 g)	Abs/2	Abs/12	Abs/2	–
Pepper				
	At harvesting		After packaging	After transportation
Total coliform (CFU/g)	1/3	4/4	2/2	2/2
E. coli (CFU/g)	0/3	0/4	0/2	0/2
Salmonella (pre/abs in 25 g)	Abs/3	Abs/4	Abs/2	Abs/2

packaging and transportation. Accordingly and to eliminate any contamination originated by farm workers during harvesting, 8 samples of zucchini were harvested by project staff and examined; contamination was detected in 4 of the 8. Results for cabbage and bell pepper have shown evidence for contamination with total coliform at harvesting—whether collected by farm workers or by project staff- and after packaging. These results demonstrate clearly that irrigation water is not the only source of contamination. In the case of Mafraq farm, it is expected that the application of manure is the source of contamination as supported by results presented by Oliveira et al. (2012).

3.2.2 Results from the Second Experiment

Tomato fruits were collected after one and two days of last irrigation, for each type of treatment. One to two samples were collected from each plot and each sample consisted of 10–12 tomato fruits. Samples collected included fruits collected from the lower (approximately 30 cm above ground), middle, and upper parts of the plant. Moreover and during the sampling event that took place after two days of last irrigation, two samples from tomato fruits irrigated with secondary effluent and touching the ground (i.e. the mulch) were collected. Results (Tables 10 and 11) have shown that high growing cultivation, use of drip irrigation, use of mulch and allowing a pathogen die off period of at least one day, have led to microbiologically safe crop, even for the fruits touching the ground.

For lettuce, samples were collected after two days of irrigation. Results (Table 12) showed total and *E. coli* levels of less than 10 CFU/g for all tested samples. Indicating that use of sub-surface drip irrigation and allowing a pathogen die off period of two days results in microbiologically safe crop.

Table 10 Microbiological quality of tomatoes sampled after one day of last irrigation

	Fresh water	Secondary effluent	Disinfected effluent
Total coliform (CFU/g)	0/3	0/7	0/7[a]
E. coli (CFU/g)	0/3	0/7	0/7
Salmonella (pre/abs in 25 g)	Abs/3	Abs/7	Abs/7

[a]Presented results are the number of samples showing total coliform or *E. coli* count above 10 CFU/g to total number of tested samples

Table 11 Microbiological quality of tomato fruits sampled after two days of last irrigation event

	Fresh water	Secondary effluent	Disinfected effluent
Total coliform (CFU/g)	0/5	0/6	0/6
E. coli (CFU/g)	0/5	0/6	0/6
Salmonella (pre/abs in 25 g)	Abs/5	Abs/6	Abs/6

Table 12 Microbiological quality of lettuce samples after two days of last irrigation event

	Fresh water	Secondary effluent	Disinfected effluent
Total coliform (CFU/g)	0/3	0/3	0/3
E. coli (CFU/g)	0/3	0/3	0/3
Salmonella (pre/abs in 25 g)	Abs/3	Abs/3	Abs/3

4 Identification and Prioritization of Hazards

Hazard can be any stressor that may cause harm to environment, human and/or properties. It is any biological, chemical, physical or radiological agent that has potential to cause harm. Hazardous event is an incident or situation that can lead to presence of a hazard (what can happen and how). Hazard identification is the process of determining stressor that may cause an increase in incidence of specific adverse health or environmental effects. Risk is the probability, which is the likelihood of identified hazards causing harm in exposed populations in specified timeframe, including magnitude of that harm and/or consequences. As seen in the results presented in the previous section, risks associated with wastewater irrigation cannot be completely separated from consequent farming practices particularly in the Jordanian context. Farming practices in Jordan are influenced by many factors that could affect a given situation. The main factors are:

- Weather conditions and season variation
- Fertilizers and pesticides application (practices and timing)
- Irrigation water quality
- Receiving and storage practices
- Sanitation and hygiene
- Handling of produce; and
- Applied irrigation system.

Hazards associated with agriculture practices due to wastewater irrigation or pesticides/fertilizers use are identified below.

4.1 Hazards Due to Wastewater Irrigation

Many researchers have evaluated negative health and environmental risks of treated wastewater irrigation. Carr et al. (2011) indicated that use of wastewater for irrigation in Jordan has the capacity to affect soil in a detrimental manner, and the effect of water on soils can be managed through application of suitable on-farm strategies. Wastewater can meet 75% of fertilizer requirements of typical farm in Jordan (Carr et al. 2011). However, excess nutrients can reduce productivity, depending upon crop. List of main hazards associated with reclaimed water irrigation (Kalavrouziotis et al. 2008; Kazmia et al. 2008; Feldlite et al. 2008; Khan and Hanjra 2008; Walker and Lin 2008; Li et al. 2009) are:

- Pathogens that can survive long enough in the environment to be transmitted to people and become serious health threat.
- Reclaimed water may lead to heavy metal transport to crops (in case of industrial wastewater).

- Nutrients imbalance may cause toxicity and adverse effects on crop yield. Humans are subject to nitrate toxicity, with infants being especially vulnerable to methemoglobinemia due to nitrate metabolizing.
- Reclaimed water has potential to induce salinity and may reduce crop production.
- Reclaimed water may leach through soil profile thus affecting the quality of groundwater (nitrate and pathogenic contamination).
- Reclaimed water may cause Irrigation system problems (e.g. clogging of drip irrigation system).

Many researchers have also identified and discussed the benefits and risks of wastewater irrigation, biophysical and socioeconomic aspects, environmental health and governance issues (Hanjra et al. 2011, 2012; Hussain et al. 2002). Hanjra et al. (2012) discussed the limitations of using wastewater for irrigation. The limitations they identified include: nutrient management, crops choice, soil properties, irrigation methods, health risk regulations, land and water rights and public education and awareness. On the other hand, irrigation with wastewater can also reduce water footprint and energy footprint of food production, earn carbon credits and potentially contribute to climate change adaptation and mitigation.

Carr et al. (2011) indicated that drip irrigation emitters became clogged due to suspended solids, mineral precipitation or algal growth as reported by farmers in Jordan. However, this does not exclude fresh water irrigation since acceptable limit of nitrate in Jordan, which is the main compound enhancing algal growth, is 50 mg/l. Also, it was indicated that effectiveness of pesticides was reduced by high pH of wastewater (Carr et al. 2011). There is a need to open up discussions and raise awareness about realities of water reuse for more efficient and productive use of reclaimed water.

4.2 Hazards from Pesticides and Fertilizers

Evidence indicates that pesticides contain chemicals that pose potential risk to humans and other forms of life and unwanted side effects to the environment (Igbedioh 1991; Forget 1993). Relevant hazard is mainly related to pesticide impact through residue in agricultural products. Analysis on crops has shown that when pesticide residue in produce exceeds maximum allowable limit, serious effect on human and animal life may occur. Additionally, pesticides can be toxic to a host of other organisms including birds, fish, beneficial insects, and non-target plants. In a detailed safety plan a list of pesticides should be prepared according to locally produced pesticides and those imported. Analysis performed by Ministry of Environment (MoE) and Ministry of Agriculture (MoA) should be studied before conclusions are withdrawn to Jordanian agricultural practices. In any case, pesticides side effects (risks) can be extended and not limited to the following:

- Some pesticides residues are known to be carcinogenic
- Surface water and groundwater contamination
- Effects on soil salinity and fertility
- Contamination of air, soil, and non-target vegetation, and
- Non-target organisms (like beneficial bacteria) can be endangered.

Regarding fertilizers, research results indicate that hazardous constituents in most fertilizers generally do not pose harm to human health or the environment (EPA 1999). However; results presented in this study indicates that un-composted manure could be a serious source of contamination as supported by results presented by Oliveira et al. (2012), and could impose human health risk.

4.3 Prioritizing Hazards

Protecting people, property, and environment from hazards is a priority. However, the constraints of time and funds preclude giving immediate attention to each hazard that may exist. Therefore, it is crucial to decide which hazards should be dealt with most urgently and which should be dealt with later or not at all. Determining which hazards to target for management is called "hazard prioritization". There is a number of ways to prioritize hazards. For this exercise the FEMA model developed by the Federal Emergency Management Agency (FEMA) of the United States was employed. In FEMA model, each hazard is rated individually using a number of quantitative criteria, and individually given a numerical score. Since FEMA model judges each hazard individually in a numerical manner, it may provide more satisfying results than other available models. In prioritizing hazards there is no "right" answer, and there will be a number of hazards that are considered to be more serious than others. The four main criteria used by FEMA evaluation and scoring system are:

- **History**: If a certain type of emergency has occurred in the past, it is known that there were sufficient hazardous conditions and vulnerability to cause event.
- **Vulnerability**: This criterion determines number of people and value of property that may be vulnerable based on some factors like vulnerable group, population densities, location of population groups, location and value of property, and location of vital facilities, e.g. hospitals.
- **Maximum threat**: This is essentially the worst-case scenario that assumes the most possible serious event and greatest impact. It is expressed in terms of human casualties and property loss.
- **Probability**: It is likelihood of an event occurring, expressed in terms of chances per year. Since some hazards are without historical precedent, an analysis of both history and probability is necessary.

FEMA Evaluation of criterion is shown in Table 13. Depending on the severity evaluation result is categorized as either low or medium or high.

Table 13 The FEMA evaluation system

Criteria			Evaluation
History: whether an emergency event has occurred:		<2 times in 100 years	Low
		2–3 times in 100 years	Medium
		>3 times in 100 years	High
Vulnerability:	of people	Up to 1%	Low
		>1–10%	Medium
		>10%	High
	of property	Up to 1%	Low
		>1–10%	Medium
		>10%	High
Maximum threat: area of the community affected		5%	Low
		>5–25%	Medium
		>25%	High
Probability: chances per year of an emergency		<1 in 1000	Low
		1 in 1000-1 in 10	Medium
		>1 in 10	High

A score is assigned for each evaluation: Low is valuated at 1Point, Medium at 5 Points and High at 10 points. Some criteria have been determined as more important than others. Therefore, the following weightings have been established: History weighted ×2, Vulnerability weighted ×5, Maximum threat weighted ×10, and Probability weighted ×7. Multiplying score by weighting, then adding the four results provides a composite score for each hazard. FEMA model suggests a threshold of 100 points. All hazards that total more than 100 points may receive higher priority in emergency preparedness. Hazards totaling less than 100 points, while receiving a lower priority, should still be considered. This process should be repeated for all identified hazards and for a range of scenarios with the same hazard.

The hazards associated with the use of wastewater in agriculture are listed in Table 14. The results given when FEMA model was applied to rank these hazards are shown in Table 15. These results indicate that the greatest hazard at farm level lies in presence and use of pesticides as they can be found as residue in agricultural products and they may be toxic. The second danger comes from contamination caused by pathogens found in wastewater or the use of un-composted manure.

The groups that are most affected by the hazards associated with pathogenic contamination are farmers and their families and then the consumers. However, it should be indicated that farmers could also be affected indirectly as these hazards may affect family income, thus giving a negative impact on the education of children and medical coverage. In a detailed safety plan, groups affected should be clearly identified and managed.

Table 14 Hazards associated with the reuse of reclaimed water in agriculture

Hazard No.	Hazard description and associated route
1	Nutrients imbalance; nutrient oversupply or deficiency may cause toxicity and impose adverse effects on crop yield
2	Accumulation of dissolved solids may reduce crop production
3	Heavy metal in industrial WW may contaminate the crop
4	Reclaimed water may leach or percolate through the soil profile thus affecting the quality of groundwater
5	Pathogens can be transmitted to people and become serious health threat
6	Pesticides pose a potential risk to humans and other forms of life and unwanted side effects to the environment (direct contact)
7	Pesticides residue in agricultural products
8	Pesticides may be toxic to other organisms including birds, fish, etc.)
9	Pesticides may lead to surface water and groundwater contamination
10	Pesticides may have effects on soil fertility
11	Crop contamination from un-composted manure

Table 15 Hazards ranking for reclaimed water agricultural irrigation and farming practices in Jordan

Hazard No.	History (×2)		Vulnerability (×5)		Maximum threat (×10)		Probability (×7)		Total
1	Low	2	High	50	Medium	50	Low	7	102
2	Medium	10	Low	5	Low	10	Medium	35	60
3	Medium	10	Medium	25	Low	10	Medium	35	80
4	Low	2	Low	5	Low	10	Low	7	24
5	Low	2	Low	5	High	10	Low	7	24
6	Low	2	High	50	High	100	Low	7	159
7	High	20	Medium	25	Medium	50	Low	7	102
8	High	20	Medium	25	High	100	Medium	35	180
9	High	20	Medium	25	High	100	Medium	35	180
10	High	20	Medium	25	Medium	50	Low	7	102
11	Medium	10	Medium	25	Medium	50	Low	7	92
12	High	20	Medium	25	High	100	Low	7	152

4.4 Hazards Management

As discussed earlier, pesticide-residues and pathogenic contamination originated from either wastewater or manure fertilizers are the hazards that can be of primary concern. Hazards related to pathogens are so far controlled as shown in Fig. 5. The figure identifies priority hazards together with their sources and the applied control measures. The process of controlling the pesticide-residues is not clear so far and the role of MoA in such control seems absent—indicated as question mark in Fig. 5. Control measures are currently applied and assured by WAJ and MoH. Limitation of the crop types is the only control measure that is applied thus far. Although the approach is successful in controlling pathogens, it is still limited particularly when the full potential of wastewater use is considered. Referring to the results of the experiments discussed earlier and other literature (WHO 2015a, b), it is clear that health protection can still be achieved even with less restrictions on irrigation water quality. This is particularly true for irrigating vegetables that can be eaten raw. Higher revenues expected from the vegetables eaten-raw makes a strong case to improve the flexibility in the treated wastewater use options (Majdalawi 2003). Based on this analogy and the results from the experiments (WHO, 2015a, b) combined with WHO guidelines (WHO 2006), the pan outlined in Fig. 6 had been proposed.

5 Proposed Framework for Implementation of SSPs

Treated wastewater use in agriculture had received attention in Jordan since 1978 when the first wastewater reuse policy was established. Policy was further developed in 1998 considering wastewater as part of the water budget with priority given to agricultural irrigation. Latest policy that was issued under "water for life" theme emphasized the importance of full utilization of wastewater. Existing policies are indeed encouraging and create good environment for establishing an optimum showcase for reclaimed water management.

Laws of the Ministry of Health (MoH), MoA, Water Authority of Jordan (WAJ), and MoE are controlling reclaimed water use for different purposes. Obviously, there are some overlaps between different bodies. Coordinated actions need to be

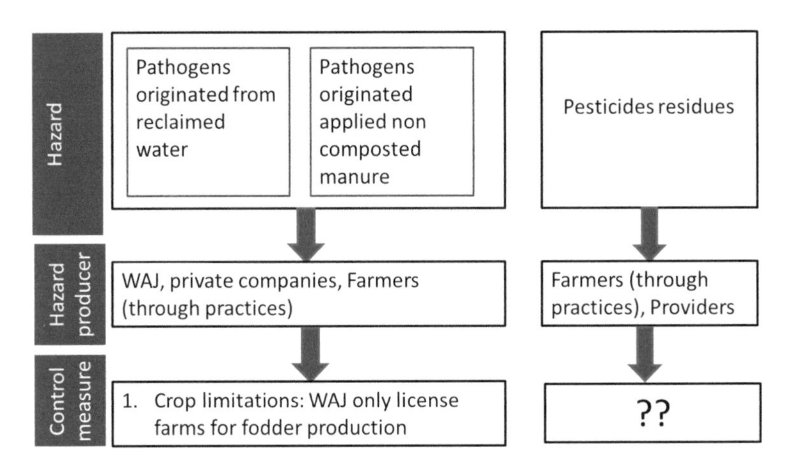

Fig. 5 Current hazard management of direct wastewater use in Jordan

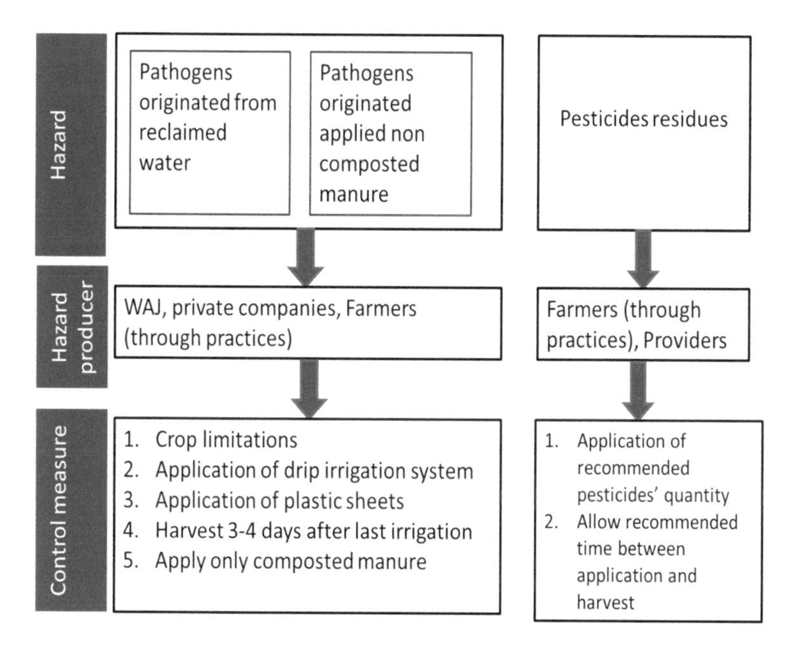

Fig. 6 Proposed control measures for priority hazards

established and better definitions and distribution of tasks are warranted. For instance, dialectic still exists between MoA and MoH about the responsibility for controlling and assuring quality of irrigated crops. Although MoA is monitoring imported crops for pesticides residues, reluctance exists for controlling local non-processed food. At the same time, MoH argues that crop quality control is part of the MoA mandate. As a matter of fact, the following articles are included in MoA's law No. 44/2002:

- Article 5B, states that the MoA contributes with competent authority in preparation and application of sanitary and phytosanitary measures to ensure prevention and transmission of disease or harm to humans through plant and animal and agricultural inputs without prejudice to any relevant authorities to examine food.
- Article 7B, states that the MoA should take sanitary and phytosanitary measures necessary to achieve proper protection of human and animal health in Jordan against the risks arising from additives or contaminants, toxins or disease-causing organisms in agricultural products or agricultural production inputs.
- Article 8, states that the MoA shall perform procedures according to the instructions issued by the Minister, which are necessary to ensure conformity of agricultural products and agricultural inputs with health and technical conditions, including sampling, inspection and control procedures.

As for MoH, the following related articles in amended law of food control for year 2003 are included. The amendment is read together with law No. 79 for year 2001:

- Article 2 defines food as any material intended for human consumption whether raw material or semi-processed or manufactured, including drinks, pickles and condiments, chewing gum and any substance used in food manufacturing, processing and treatment except for cosmetics, tobacco, drugs and drinking water.
- Article 3 states that: Subjected to the provisions of Agriculture Law in force, the institution is the only agency authorized to oversee health and food control, including suitability for human consumption in all stages of trading, whether locally produced or imported in coordination with any official related entity if the Director-General sees the need for such coordination.
- Article 11 A, states that the MoH takes in accordance with instructions issued by the Minister, necessary measures to ensure that the food and health conditions are met or health measures, including sampling, inspection and control procedures are implemented.

There are certainly overlaps in responsibilities assigned for each ministry, which requires careful coordination between both for adequate control of produce. In addition to existing overlap and lack of coordination, it is expected that lack of capacities at both ministries are behind reluctance to take decision on responsibilities to control locally produced crops. Institutional, infrastructure and human capacities at both ministries require improvement. Recently, and through Jordanian component of WHO/UNDP/GEF project on adaptation to climate change, capacities of Jordan Food and Drug Administration (JFDA) was built for measuring pathogenic contamination of crops. On the other hand, MoA has the capacity to monitor pesticides residuals in crops, whether imported or locally produced.

5.1 Scenario Analysis

Obviously, implementation of JS 1766/2014 would require additional efforts to establish a clear set up where full agricultural process rather than reclaimed water quality is controlled. Accordingly, the following two scenarios have been suggested:

5.1.1 Scenario I

This scenario proposes establishing a unit at MoH, which has the following responsibilities:

1. Issuing licenses to the farmers applying for reclaimed water use, whether direct or indirect use.
2. Control and monitor the produce with respect to pesticide-residues and pathogenic contamination. Capacity of JFDA should be built accordingly. Samples have to be collected from farms directly since number of WWTPs is limited, even when considering the ones planned for near future.
3. Take corrective actions if tested samples fail to meet produce quality according to recognized standards.

Licensing the use of treated wastewater in unrestricted irrigation would require the users to be trained and certified for good agricultural practices recommended by MoA. The MoA should also approve amounts and sources of applied fertilizers and pesticides. WAJ has to provide data related to irrigation water quality to MoA and MoH on a yearly basis. With respect to un-composted manure, Jordan currently has around 5000 composters. Controlling such a high number of composters presents a challenge to MoA. One solution can be to establish associations and make the associations responsible for quality control of the end product. This will automatically put wastewater use for agricultural production in a wider context where responsibilities of MoA has to be activated for controlling all inputs to agricultural fields including fertilizers, pesticides, and irrigation water quality.

5.1.2 Scenario II

Responsibilities are shared between an established unit at the MoA and another established unit at the JFDA. The unit at the MoA shall be established under direct responsibility of assistant secretary general for plant wealth (referring to organizational chart of MoA). The unit shall have the following responsibilities:

1. Issuing licenses for farmers applying for reclaimed water use, whether direct or indirect is are intended. Licensing should be done based on the best agricultural practices and training certificates to the farmers.
2. Control and monitor the produce with respect to pesticide-residues. Samples have to be collected directly from farms.
3. Control and monitor the agricultural practices at the farm and approve their compatibility for licensing purposes.
4. Take corrective actions if tested samples fail to meet produce quality according to recognized standards or when agricultural practices do not meet licensed practices.

In both scenarios, activation of MoA's role through the agricultural extension unit, in raising awareness between farmers on good agricultural practices is crucial. MoA can establish training programs necessary for proper implementation of JS 1766/2014. A second possibility would be certifying the private sector to perform necessary training programs.

Table 16 Advantages and disadvantages foreseen for location of suggested established unit(s) responsible for reclaimed water reuse management

	Advantages	Disadvantages
Scenario I	• Responsibilities of monitoring and control are concentrated in one unit	• Needs significant capacity building of JFDA • Limits the role of MoA
Scenario II	• Capacities of both JFDA and MoA are utilized • Fits better into legal role assigned for each ministry	• Needs higher level of coordination between MoA and MoH • Samples has to be collected twice from each farm (resources loss)

Advantages and disadvantages of both scenarios are shown in Table 16. While first scenario concentrates responsibility in one unit located at JFDA, it limits the role of MoA according to its law No. 44/2002. On the other hand, distributing responsibilities between JFDA and MoH would utilize resources and capacities available at each body. However, it also has some limitations related to higher level of required coordination and some limited duplication.

5.2 Stakeholder Consultation

Two round table discussions were held aiming at consulting the decision makers on required arrangements for implementing the relevant local guidelines, i.e. JS1766/2014 (see Chap. 5 for more information) and the suggested scenarios for the framework developed for SSP implementation. Decision makers represented main governmental bodies involved in the proposed framework. Secretary Generals (SGs) or assistants of SGs of the Ministry of Water and Irrigation, WAJ, JVA, and MoA attended round table discussions. Major outcomes are summarized as the follows:

1. MoH shares WHO understanding of the importance of controlling the whole chain in order to obtain target agricultural produce quality.
2. Cost associated with wastewater treatment, especially for tertiary treatment, can be reduced by implementing concepts presented by WHO (2006) guidelines and the following adapted JS1766/2014 guidelines. Consequently, budgets can be secured for additional sanitation services in Jordan.
3. A follow-up is extremely important to eventually develop a detailed SSP. This is of extreme importance when it comes to trade treaties and exports.
4. Media and the society have to be addressed in order to improve awareness about wastewater use in preparation for the implementation of JS1766/2014.
5. Efforts exist at MoA with respect to produce a tracking system, especially with those planned for export. There is a need to build on experience of MoA in this regard. Farmers interested in joining the system are required to fill a special application form at MoA or any of directorates that are run by the ministry.

MoA submits the application to the certification body. The certification body is to verify the documents submitted with the application. Application is then approved or deficiencies have to be addressed. Certification body also has the mandate to do inspection of operators (farmers) and to verify that they are complying with instructions set forth in this body system. Certification body shall provide MoA with names of operators who fulfilled conditions or who do not meet conditions. Certification body grants qualified operators the "instructions certificate" and informs MoA. Operators who fulfilled conditions of accession and committed themselves to apply these system instructions are granted the right to use Jordanian quality mark on their products adopted in this system. Quality system instructions related to traceability were issued on 2012 based on articles 3, 4, 8, and article 11 of agriculture law number 44/2002. The system was developed for selected products namely tomato, cucumber and dates. Though the system is still not obligatory, it is a step towards controlling quality of produce that can be used later for products grown with different water qualities including reclaimed water.

6. There was an agreement that farmers associations should play a main role in the implementation of SSPs. There is a need to upgrade the role of cooperation to activate guidelines of good agricultural practices. This would also encourage development of good documentation for all activates at farm level and their outputs.
7. There was an agreement on the need for developing the capacity of JFDA to take its role in testing fresh produce. Moreover, there exists a need to develop local guidelines for production of fresh produce.

6 Concluding Remarks

Facts presented in the chapter shows that irrigation water is only one element in a bigger matrix impacting quality of agricultural produce. Other elements can be even more serious in the Jordanian context and are related to pesticide-residues and manure applications, which have to be controlled. It should be noted that, out of the many steps involved in the wastewater irrigated agriculture, the study was limited to the agricultural field. However, end product reaching consumer should be of main concern. Shipping and handling of crops can also be an additional source of pathogenic contamination. This would be an additional dimension that locates irrigation water in a wider context. When borders of the study expand, samples have to be collected from the markets as well. In such case, traceability of the produce is crucial to find the source if/when contamination is discovered.

The studies explained in this chapter also revealed that the irrigation water is only one element affecting produce quality. While it may present a risk with respect to pathogenic contamination, manure fertilizer can be a main source that did not receive required attention and control so far. It was clearly shown that even applying drinking water quality for irrigation, a produce meeting recognized standards with respect to pathogens is not guaranteed at farm level. Particularly, non- composted manure was responsible for such contamination.

Another important point is the role of agricultural extension in raising awareness between farmers on quality of inputs applied in their farms. Quality of irrigation water and the quality and quantity of fertilizers and pesticides are all important factors affecting the end product. Farmers should be aware that excessive quantities of such inputs may negatively impact not only their products, but also their soil and environment. Ultimately, it will affect the economic value of their farms and products. Finally, pilot farm where traceability of the produce is implemented can be advantageous and could also serve as local as well as regional model for successful use of WHO (2006) guidelines. Some farmers in Jordan apply traceability systems in their farms and already established external market for their products. It will be wise if cooperation with such farmers is established to present a model case for quality control of wastewater use in agriculture.

Hazards identification showed that in addition to quality of irrigation water, other inputs of main concern include pesticides and manure. Produce quality can be negatively affected when such inputs are not controlled. Application of some control measures were shown to be effective in hazard management. When it comes to pathogens control, measures that have to be taken at farm level were identified in such ways that guarantee production of produce that meet recognized enforced standards.

As discussed earlier, the herein established frame is only meant to provide the foundation for SSP for agricultural use of wastewater in Jordan. Detailed SSPs are supposed to cover all the necessary elements including better description of hazards (based on surveys); groups exposed to hazards; better definitions of roles of MoE; detailed corrective actions; validation and verification of the plan as presented by SSP manual (WHO 2015b). All parallel supporting activities shall also be described.

References

Ammary, B. (2007). Wastewater reuse in Jordan: Present status and future plans. *Desalination, 211,* 164–176.

APHA. (2012). *APHA, AWWA, WEF. Standards methods for examination of water and wastewater*, 22nd ed. Washington: American Public Health Association; 1360 p. ISBN 978-087553-013-0. http://www.standardmethods.org/.

Arab Countries Water Utilities Association (ACWUA). (2011). Safe use of treated wastewater in agriculture: Jordan case study, Prepared by Eng. Nayef Seder (JVA) and Eng. Sameer Abdel-Jabbar (GIZ), Amman, Jordan.

Batarseh, M., & Tarawneh, R. (2013). Multiresidue analysis of pesticides in agriculture soil from Jordan Valley. *Jordan Journal of Chemistry, 8*(3), 152–168.

Carr, G., Potter, R. B., & Nortcliff, S. (2011). Water reuse for irrigation in Jordan: Perceptions of water quality among farmers. *Agricultural Water Management, 98,* 847–854.

Davison, A., Howard, G., Stevens, M., Callan, P., Fewtrell, L., & Deere, D., et al. (2005). Water safety plans: Managing drinking-water quality from catchment to consumer. WHO/SDE/WSH/05.06. http://www.who.int/water_sanitation_health/dwq/wsp170805.pdf. Accessed on February 4, 2017.

EPA, U.S. Environmental Protection Agency. (1999). Estimating risk from contaminants contained in agricultural fertilizers. Prepared by Office of Solid Waste and Center for Environmental Analysis Research Triangle Institute.

Feldlite, M., Juanicó, M., Karplus, I., & Milstein, A. (2008). Towards a safe standard for heavy metals in reclaimed water used for fish aquaculture. *Aquaculture, 284*, 115–126.

Forget, G. (1993). Balancing the need for pesticides with the risk to human health. In G. Forget, T. Goodman, & A. de Villiers (Eds.), *Impact of pesticide use on health in developing countries* (p. 2). Ottawa: IDRC.

Hanjra, M. A., Raschid, L., Zhang, F., & Blackwell, J. (2011). Extending the framework for the economic valuation of the impacts of wastewater management in an age of climate change. *Environmental Management* 1–14.

Hanjra, M., Blackwell, J., Carr, G., Zhang, F., & Jackson, T. (2012). Wastewater irrigation and environmental health: Implications for water governance and public policy. *International Journal of Hygiene and Environmental Health, 215*, 255–269.

Hussain, I., Hanjra, M. A., Raschid, L., Marikar, F., & Van Der Hoek, W. (2002). *Wastewater use in agriculture: Review of impacts and methodological issues in valuing impacts with an extended list of bibliographical references*. Working Paper 37, International Water Management Institute, Colombo, Sri Lanka.

Igbedioh, S. (1991). Effects of agricultural pesticides on humans, animals and higher plants in developing countries. *Archives of Environmental Health, 46,* 218.

Jordan Valley Authority (JVA) and Ministry of Agriculture (MoA). (2010). Annual Report, JVA and MoA, Amman, Jordan.

Kalavrouziotis, I. K., Robolas, P., Koukoulakis, P. H., & Papadopoulos, A. H. (2008). Effects of municipal reclaimed wastewater on the macro- and micro-elements status of soil and of *Brassica oleracea* var. Italica, and *B. oleracea* var. Gemmifera. *Agricultural Water Management, 95*, 419–426.

Kazmia, A. A., Tyagia, K., Trivedi, R. C., & Kumar, A. (2008). Coliforms removal in full scale activated sludge plants in India. *Journal of Environmental Management, 87*, 415–419.

Khan, S., & Hanjra, M. A. (2008). Sustainable land and water management policies and practices: A pathway to environmental sustainability in large irrigation systems. *Land Degradation and Development, 19*(3), 469–487.

Li, P., Wang, X., Allinson, G., Li, X., & Xiong, X. (2009). Risk assessment of heavy metals in soil previously irrigated with industrial wastewater in Shenyang, China. *Journal of Hazardous Materials, 161,* 516–521.

Majdalawi, M. (2003). Socio-economic and environmental impacts of the re-use of water in agriculture in Jordan. Farming systems and resources economics in the tropics No 51. Dissertation. Hohenheim University, Stuttgart, Germany.

Ministry of Environment MoE. (2016). Second environmental status report.

MWI. (2016). National Water Strategy of Jordan 2016–2025, Ministry of Water and Irrigation publication.

Murtaza, G., Ghafoor, A., Qadir, M., Owens, G., Aziz, M. A., Zia, M. H., et al. (2010). Disposal and use of sewage on agricultural lands in Pakistan: A review. *Pedosphere, 20*(1), 23–34.

Oliveira, M., Vinas, I., Usall, J., Anguera, M., & Abadias, M. (2012). Presence and survival of *Escherichia coli* O157:H7 on lettuce leaves and in soil treated with contaminated compost and irrigation water. *International Journal of Food Microbiology, 156*(2), 133–140.

Walker, C., & Lin, H. S. (2008). Soil property changes after four decades of wastewater irrigation: A landscape perspective. *Catena, 73*, 63–74.

Water Authority of Jordan (WAJ). (2013). Agreements with farmers for purposes of reusing treated wastewater in irrigation. Technical report,, Water Reuse and Environment Unit, WAJ, Amman, Jordan.

WHO. (1989). Health guidelines for the use of wastewater in agriculture and aquaculture. Technical report series 778. ISBN 92 4 1207787.

WHO. (2006). *Guidelines for the safe use of wastewater, excreta and grey water*. Geneva: World Health Organization.
WHO. (2015a). Stakeholder analysis and pilot study for safe use of treated wastewater in agriculture. Final report: Framework for sanitation safety plan: Reclaimed water use in agriculture. Report under contract number EM-CEHA-2014-APW-016.
WHO. (2015b). Sanitation safety planning. Manual for safe use and disposal of wastewater, greywater and excreta. ISBN 978 92 4 154924 0.

Public Acceptance of Wastewater Use in Agriculture: Tunisian Experience

Olfa Mahjoub, Amel Jemai, Najet Gharbi, Awatef Messai Arbi and Souad Dekhil

Abstract Use of wastewater in agriculture has become commonplace in many countries where fresh water scarcity is already a reality. Public acceptance plays a key role in such projects and due attention should be paid before, during, and after the project implementation. This chapter aims at: (i) giving an overview on the status of the agricultural use of treated wastewater in Tunisia and the main hurdles hampering its progress and, (ii) showcasing one of the most successful irrigated areas in the region of Ouardanine, to determine the factors that have made it prosperity, while focusing on the social dimension and the perception of end-users. Aspects related to education, knowledge, risk perception, culture, regulation, and communication need to be seriously addressed for a more viable and efficient use of wastewater in agriculture. The use of wastewater in Ouardanine has flourished, exceptionally well and seven factors were identified as the drivers of this success. The perceived financial benefit was ranked first while the lack of fresh water resources in the region was the second. Environmental awareness and the impact of non-reuse option in the region should be underscored. While the acceptance of farmers was high, the reluctance of consumers was still impeding market share; more relaxed regulation together with good practices is suggested as option to improve the situation.

Keywords Public acceptance · Agriculture · Awareness · Ouardanine
Wastewater irrigation · Water quality · Financial feasibility

O. Mahjoub (✉) · A. Jemai
National Research Institute for Rural Engineering, Water, and Forestry (INRGREF), 2080 Ariana, Tunisia
e-mail: mahjoub.olfa@iresa.agrinet.tn; olfama@gmail.com

N. Gharbi · S. Dekhil
Department of Rural Engineering and Water Exploitation, Ministry of Agriculture, Hydraulic Resources and Fishery (DG/GREE, MARHP), Tunis, Tunisia

A. M. Arbi
Department of Environment and Quality of Life (DGEQV), Ministry of Local Affaires and Sustainable Development, Tunis, Tunisia

H. Hettiarachchi and R. Ardakanian (eds.), *Safe Use of Wastewater in Agriculture*,
https://doi.org/10.1007/978-3-319-74268-7_7

1 Introduction

The new developments in the wastewater treatment technologies have improved its potential as anon-conventional water resource to counterbalance water shortage and supply limitations worldwide. Social and cultural acceptance are crucial for the successful implementation of wastewater use in agriculture. Some wastewater use schemes have even been halted by the lack of public acceptance. One such example comes from the area of Cebala (Borj Touil city) in Tunisia, where poor public acceptance (combined with other reasons) caused the wastewater irrigated area to shrink from 3200 to 190 ha. Negative public attitude against treated wastewater use has been a major inhibitor, especially in agriculture. Some may argue that the public acceptance of wastewater use is not an obstacle by itself; it is rather the pessimistic perception that causes reluctance to use (Baumann 1983).

Acceptance of wastewater use can be categorized into two groups. First category includes those who use of treated wastewater with the knowledge of its potential risks. The second category is where the risk perceptions are low and public health is potentially challenged by the common use of untreated, partially treated or diluted wastewater. In both situations, there is a need to gain trust (Drechsel et al. 2015). Based on the studies carried out during the 1960s and 1970s, acceptance varied according to potential use and was embedded in cognitive factors. Awareness on water supply, treatment, distribution of wastewater, and income were believed to influence the perception of reuse. Age, political affiliation and attitudes to local government were considered peripheral factors while price and psychological factors were of little influence on the level of acceptance (Baumann 1983). The cost of wastewater treatment and thereby the cost of reuse does represent great concern (Buyukkamacia and Alkan 2013). Strict regulations may compromise the economic viability of wastewater use schemes in regions where planned reuse is already being practiced (Grundman and Maas 2017). Health risks associated with close physical contact with wastewater may not be accepted by the public (Buyukkamacia and Alkan 2013).

Success stories in agricultural applications of treated wastewater have flourished worldwide. However, successful cases of reuse in developing countries do not look alike. In fact, treatment can be limited to secondary stage and quality of treated effluents may be below the level required. In addition, regulations are rarely set and enforced. This chapter describes the important aspects of public acceptance of wastewater use in agriculture through an interesting example from the Ouardanine region in Tunisia. A great deal of it is based on information collected during interviews realized with farmers in the region, either as a structured survey or as free and open discussion. It also highlights the main outcomes of a study commissioned for the Ministry of Environment and Sustainable Development to establish a national strategy for sensitizing the current and the potential future users of wastewater to good practices of safe agricultural use of wastewater.

This chapter is aimed to be a critical review of the current situation of treated wastewater use in agriculture in Tunisia and an analysis of its public acceptance.

More particularly, it is meant to highlight the main outcomes of an extensive survey carried out in the irrigated area of Ouardanine to emphasize on the factors behind social and cultural acceptance that have contributed to the progress and rapid extension of the wastewater irrigation in this area.

2 Wastewater Use in Tunisia

Wastewater is recognized worldwide as a reliable water resource because of its increasing volume and assurance of availability all year round. In Tunisia, the volume of wastewater has evolved substantially since 1975. Current, wastewater in Tunisia is treated secondarily in 113 wastewater treatment plants (WWTP). The produced volume of 243 MCM (ONAS 2015) represents about 5% of the total water resources available in the country (Fig. 1) and is expected to double by the year 2020 because of population growth and economic activities development (DGGREE 2016).

2.1 Wastewater Irrigation

Like many other countries the largest user of freshwater in Tunisia is the agricultural sector. It consumes around 80% of the available fresh water resources to grow food for more than 11 million people (ITES 2014). According to the forecasts, agriculture will need to feed 2 million more Tunisians by 2030 which will increase the water demand close to 2760 million m^3/year (MCM/year). By that time, Tunisia will be suffering from acute water scarcity with a 370 m^3/year/person share of water (ITES 2014). This will push the agricultural sector to rely on other water sources i.e. non-conventional water.

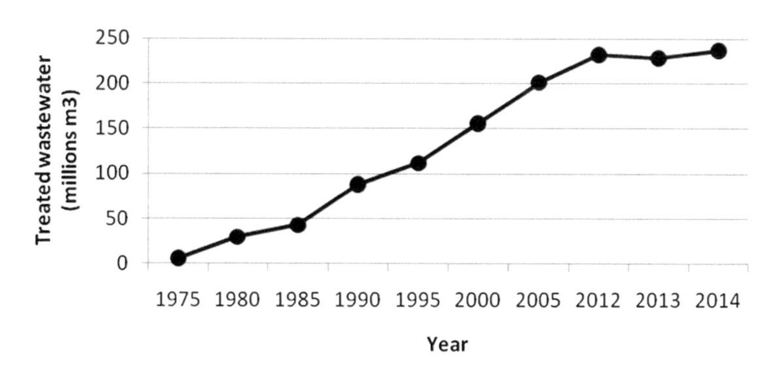

Fig. 1 Evolution of the volume of treated wastewater produced between 1975 and 2014. *Source* DGGREE (2016)

Treated wastewater (TWW) is an attractive water source that can substitute conventional water especially in the agricultural sector as it can improve crop yield and thereby contribute to the economic welfare in peri-urban and rural areas. However, only 20–45% of TWW is reused for irrigation in agriculture. Wastewater used for irrigation is supplied by 23% WWTP producing about 60% of the total produced volume (DGGREE 2016). Many existing WWTP are old and several are practicing illegal discharge of industrial effluents. Number of them is have exceeded their treatment capacities by up to 150% which has resulted in the degradation of the quality of the effluents either used for irrigation or discharged in the receiving environment. Between 2011 and 2012, about 17 MCM of effluents, corresponding to 7% of the treated effluents, was used for irrigation, which represented only 42% of the total volume available for irrigation. Reused volumes of TWW reported in literature are likely to be biased and overestimated because of possible leakages and illegal reuse practices are not accounted for.

As per the irrigated area, 1200 ha of orchards were irrigated with wastewater in 1965, driven by the depletion of the aquifers in the northern part of the country. The expanding urbanization and suburbs has shrunk the area down to 400 ha. During the period 1965–1989, the agricultural reuse has been boosted by the setting of political, regulatory, and institutional frameworks. In addition, there was a thriving scientific research on the topic that has led to a rapid expansion of the irrigated area up to 6500 ha. This period was characterized by some intense scientific research activities and many outcomes featuring the characteristics of wastewaters and their potential benefits in the agricultural sector and the subsequent impacts on the environment (Bahri 1998; Rejeb 1990; Trad-Rais 1988). The establishment of the national standards related to the agricultural reuse in 1989 and the definition of legislative framework have contributed significantly in the registered success of this practice and the promotion of a safe reuse in agriculture (INNORPI 1989). This progress is obviously not disconnected from the global trends in the reuse observed worldwide and the setting of the guidelines (WHO 1989).

The following twenty-five years of TWW use was characterized by an evident and substantial slow-down and even a decline of the irrigated areas (Fig. 2). Nowadays, reuse is practiced over an area of around 8150 ha, representing only a 2% of the total irrigated land in Tunisia. The exploited area is about 75% of the total, estimated at 6104 ha. However, these figures vary over time and upon sources. The size of the irrigated area is lower due to the difference between what is called “equipped area” and “actual irrigated area” which may be slightly confusing, leading to overestimation. In 2012, the equipped area was around 8036 ha while the actual irrigated area was only of 2215 ha. The large difference between the two figures reflects a tremendous variation of the exploited/irrigated areas. In fact, 20% of the land equipped with distribution valves are completely abandoned (Fig. 3) due to several reasons ranging from forthcoming rehabilitation and/or extension to a deliberate and complete rejection of reuse by the farmers stemming from a range of factors that will be depicted in the following sections (DGGREE 2016).

In 2016, 28 areas equipped with valves and ready to be cultivated were distributed over 15 governorates. The area of Borj Touil, one of the largest, has

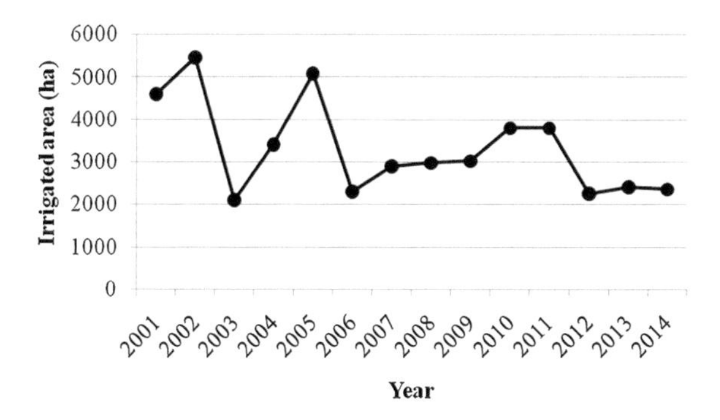

Fig. 2 Evolution of the actual total irrigated area between 2001 and 2014 in Tunisia. *Source* DGGREE (2016)

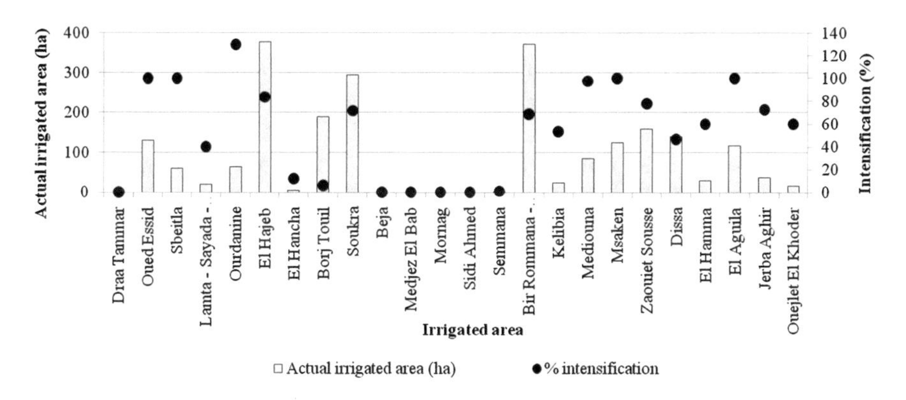

Fig. 3 Actual irrigated areas with TWW and percentage of intensification. *Source* DGEQV (2013)

witnessed a drop from 1200 to 190 ha caused mainly by the farmers' reluctance. Together with the area of Mornag (1087 ha), they account for 53% of the total area, both located in the northern part of the country well-endowed with fresh water resources (DGGREE 2016). The percentage of intensification is highly variable amongst the irrigated lands. It may reach up to 140% in the most successful areas like Ouardanine. On average, the mean value is around 30% country wide (Fig. 3) (DGGREE 2016). This also reflects the low public acceptance and the obstacles to a potentially thriving sector while the country is witnessing a critical water scarcity.

2.2 *Regulations and Institutional Issues*

The use of TWW in agricultural irrigation in Tunisia unofficially started more than 50 years ago in the region of Soukra, north of the country, driven by an urgent need to preserve citrus orchards. However, the regulations and national standards for TWW use in agriculture were not set the late 1980s. The regulations helped the government to recognize wastewater an asset in closing the gap of water demand in irrigation. The 10th National Development Plan set for the period 2003–3007 targeted 60% of the effluents produced, including all uses (aquifer recharge, agricultural irrigation, and irrigation of landscape and gulf courses), and an irrigated area expanding to 22,000 ha (Neubert and Benabdallah 2003). These objectives were very optimistic and they have never been achieved. The National Strategic Plan set for the period 2016–2020 is targeting a 50% use of TWW for various purposes (National Development Plan 2016). The new legal and regulatory framework which is more relaxed, is expected to enhance this practice.

Various obstacles have been identified by the stakeholders of the water sector including health and environment. Given the transitional political situation, transformation of these hurdles into opportunities is very challenging. The most critical issues are summarized below:

- Some old WWTPs are producing low quality of TWW. In fact, 10 irrigated areas are supplied with effluents that do not comply with the national quality standards of reuse. During the period 2009–2012, 17 WWTPs did not meet the standards of reuse for 50% of the monitored parameters and 10 were exceeding the standards for all the measured parameters (DGGREE 2016). TWW quality is also impaired by the presence of industrial effluents in the sewage network, released illegally or without preliminary treatment, due to low enforcement of discharge regulations in the sewerage network and in the water bodies.
- The government failed in applying an attractive pricing strategy. The tariff adopted for the promotion of reuse of reclaimed water in agriculture is not up-to-date. Currently, TWW is not priced attractively in a way that allows cost recovery for maintenance and for the renewal of the irrigation network. Indeed, end-users often associate the low price to a low quality.
- There is lack of institutional capacities and collaboration between the institutions in charge of water, environment, and health sectors at national, regional and local levels. The absence of outreach to the farmers is due to the limited available capacities within the institutions. Besides, the scientific research outcomes are seldom translated into applicable solutions.
- Some areas irrigated with wastewater in the northern part of the country, do have access to conventional water sources such as rain water). The availability of conventional water on 53% of the total irrigated land is preventing farmers from bearing in mind the advantages of reuse, including the augmentation of the available water resources and the gain of fertilizers. However, it should be noted that area does only correspond to two irrigated areas out of the existing twenty-eight.
- Public acceptance has been overlooked for decades by the national strategies despite that social and cultural believes are the backbone of reuse projects. Multitude of factors is contributing to the negative perception of TWW used for irrigation, which still represent an obstacle. These factors will be depicted in detail in the next sections.

3 Social Acceptance of Wastewater Use in Tunisia

Public acceptance is the corner stone of any project of wastewater use in agriculture either it is using raw or TWW. Studies carried out worldwide have identified gaps to be dealt with in promoting different types of use, not only in agricultural irrigation. The first unique study on this topic in Tunisia aiming to address this issue at national level, was carried out during the period 2011–2013. In the former times, social aspects were more commonly dealt with on the creation of a new irrigated scheme, as part of feasibility study.

There is a multitude of determinants of social acceptance that challenges the success of a project when it has already been introduced without strategically addressing the social and cultural aspects. In United States, USEPA has funded an interdisciplinary and integrative social science study on public perception and participation in water reuse within the country (Hartley 2006). In Tunisia such social aspects have never been studied by social scientists as a standalone topic and social acceptance of water reuse has always been associated with the economics (Özerol and Günther 2005; Selmi et al. 2007; Zekri et al. 1997). Probably, the misconception was the thought that it could be more comprehensive to investigate the financial benefits perceived by the end-users, i.e. farmers. In doing so, several aspects were left behind. Also, consumers' perception of wastewater irrigated products is often overlooked; in the best case it is indirectly evaluated through the difficulties encountered by farmers in finding marketing channels. One survey reported that about 39% of farmers were not able to sell their fruits in local markets (DGEQV 2013) due to the negative public perception. Questioning consumers and collecting one's opinions about this topic is likely so sensitive that performing a survey is feared to induce a negative reaction and develop reluctance in lack of a strategic study (DGEQV 2013). Besides, farmers have always been considered the user of TWW but never questioned about their attitude as consumers of their own produce.

3.1 *Knowledge and Education*

Throughout Tunisia, the number of farmers practicing wastewater irrigation is estimated to be 2350 and a majority of them are of a poor educational background; 18% of illiterate and 47% have only attended primary schools. A few have received university education with predominance in the northern part of the country (DGEQV 2013). These figures may not be supportive of reuse because of the well-known relationship between people's attitude and their formal education (Baumann 1983).

The Agricultural Development Group (abbreviation in French GDA), is a group of farmers' representatives operating like an NGO. It is recognized to contribute a lot in providing outreach and adopting good practices in wastewater use. It also promotes compliance with regulations in terms of irrigated crops. Yet, a little more than 50% of farmers are adherent of GDA, which explains partially the lack of knowledge about the use of wastewater in agriculture. Indeed, 70–80% of farmers in the north and the center of the country declared receiving no information prior to the implementation of reuse projects. Surprisingly, 80% seem to attribute no advantage to the reuse of wastewater, particularly as a source of nutrients. Thereby, it becomes obvious that about 75% of farmers are adding fertilizers on the top of wastewater (DGEQV 2013).

3.2 Risks Perception

In the early 80s, it was reported that in the peri-urban area of Soukra, 83% of the farmers were in direct contact with wastewater when irrigating. However, only 62% declared taking a shower after irrigating their fields, only 54% were wearing boots, 7% used gloves and partial vaccination was taken by only 6% of the farmers' population (Zekri et al. 1997). A recent comprehensive study revealed that 40–50% of the farmers considered that health risks are the main obstacle to the wastewater use. However, 70% of them still do not take vaccination, do not wear protective clothes, and do not shower after irrigation. In addition, 100% of them have not undergone any medical examinations. Examples from above studies are not unique; in fact, many others throughout the country have clearly seen that awareness is not linked to a specific practice. Hence, as of now, there have been no scientific studies published on risk perception among users of wastewater in irrigation, as farmers or the laborers who work in the fields.

3.3 Cultural and Religious Beliefs

The World Health Organization (WHO) guidelines recognize need to consider cultural and religious factors to make wastewater irrigation practices successful (WHO 2009). While cultural and religious aspects related to wastewater use could be addressed independently, there is a clear trend in combining them because religious beliefs are embedded in the culture.

In Tunisia, rejection of wastewater use based on cultural and religious beliefs has never been officially reported. This could be because of the sensitivity of the topic and the lack of skills and methodology to approach the population without inducing a negative reaction. The very first publication dealing with this topic was a comparative study between Jordan and Tunisia in which around 20% of the interviewees were reported to be against wastewater use, either restricted or unrestricted, because of religious prohibition (Abu Madi et al. 2003). More recently, it was found that globally around 33% do mind wastewater use based on cultural and religious beliefs. Based on regional distribution, rejection was more pronounced in the northern part of Tunisia where 43% of the farmers were opposing this practice. It is interesting to note that the TWW use has been recognized by the World Fatwa Management and Research Institute which is an important and respected entity in the Muslim world who say that "If water treatment restores the taste, color, and smell of unclean water to its original state, then it becomes pure and hence there is nothing wrong to use it for irrigation and other useful purposes" (INFAD 2012).

3.4 Regulatory Framework

Great majority of the existing regulations are inspired by the 1989 WHO guidelines, recognized to be very restrictive. These regulations were often seen as an obstacle to the promotion of TWW use in developing countries especially among farmers who were used to growing crops eaten raw. The new WHO guidelines published in 2006 were renowned to be more permissive by applying the concept of health-based targets and performance targets. Wastewater use in Tunisia did not benefit from this relaxation. Its application was challenging for policy makers and practitioners, indeed.

Amending national regulations is time consuming because on one hand discussions among the multitude of institutions should lead to a consensus, and on the other hand, baseline data and research outcomes were required to facilitate setting realistic threshold concentrations for parameters. In Tunisia, national regulation was deemed supportive of protecting health and the environment, but restrictive of promoting the practice. Considering the current context and the global change, the process of adjusting the regulatory and the legislative frameworks was commenced later on.

3.5 Communication

Knowledge sharing and communication are crucial factors in enhancing acceptance during implementation of a safe use of TWW in agriculture (Drechsel et al. 2015). Tunisia is facing several obstacles in this regard and the severity depends on the geographical location and the level of knowledge of the communities (DGEQV 2013). The Agricultural Extension and Training Agency (abbreviated in French as AVFA) oversees the implementation of training programs elaborated by Regional Department for Agricultural Development (abbreviated in French as CRDA), targeting technicians and farmers. It is also in charge of the preparation of educational materials (AVFA 2008). In former times, AVFA produced number of brochures exclusively focused on crops allowed practices not allowed in wastewater irrigation. These brochures were not attractive to the end-users neither in terms of language nor design.

A critical study made on the communication strategy adopted by AVFA in promoting wastewater use in agriculture revealed a weak collaboration between institutions involved in wastewater use and the absence of professionalism in the conception of the material. The language used to address the farmers included sophisticated technical terms not easily understood by them. As for the design and conception, the existing material reflected lack of creativity and esthetic aspects. The content of the material used for awareness campaigns and for extensions, did not cover the latest scientific findings, and the information included was not easily understood (DGEQV 2013).

Lack of monitoring (follow-ups and of evaluation) was also noticed, supported by the fact that only 13.4% of the farmers could recognize the logo of the AVFA in the brochures. In northern part of the country, 83% of the farmers did never have any visits from AVFA/CRDA technicians. This partially explains the high reluctance of the land owners to use wastewater in that region. In rural areas, GDAs oversee wastewater management at local level and maintenance of the irrigation network. GDAs play an important role in facilitating the implementation of impactful activities. They may facilitate disseminating success stories and good practices among farmers to increase knowledge in addition to creating awareness. However, the financial issues faced by GDAs has threatened these positive roles they can play.

To tackle these obstacles, awareness campaigns about the safe reuse of wastewater in agriculture were organized in 2013 in three pilot areas in north, center, and south of the country. The outcomes were mixed and rather showed a global trend than real facts. In fact, only 43% of the targeted individuals did participate and 20% were not satisfied. The most successful topics were related to hygiene, protective tools, and regulation. A strategy of communication based on the Theory of Planned Behavior is being considered now to be implemented during the period 2015–2019. The strategy is aimed to be revised, updated and adjusted based on the regional and local peculiarities. It encompasses the following components:

- Improvement of the communication/information system;
- Promotion of the good practices and benefits of the reuse;
- Communication for the development and mobilization of networks and partnership.
- The following topics were set as priorities (DGEQV 2013):
  - Sustainable water resources management;
  - Respect of regulation related to reuse;
  - Health and hygiene aspects related to reuse;
  - Technical aspects related to wastewater management.

4 Wastewater Irrigation in Ouardanine, Tunisia

In Tunisia, case studies that have been monitored and deemed as thriving examples are rare. One such rare example is the region of Ouardanine. The use of treated wastewater (TWW) for irrigation in Ouardanine dates back to the 1990s. Since then a sustainable agricultural practiced has been achieved through progressive adoption of good practices (Mahjoub et al. 2016).

The district of Ouardanine is located in the Center East of Tunisia, 130 km from the capital Tunis. It belongs to the governorate of Monastir (Fig. 4). The region has a semi-arid climate and has been undergoing a water deficit estimated to be around 1000 mm/year (Mahjoub et al. 2016). The current wastewater irrigated area of Ouardanine was once a large orchard composed 97% of olive trees. Rain-fed agriculture was the rule and trees were watered by a system of *meskat*, the traditional rainwater harvesting system (Mahjoub et al. 2016).

In the early 1990s the effluents from the city of Ouardanine sewerage network were not treated properly; instead they were discharged in a stream crossing the agricultural area, called Oued El Guelta which eventually become a stream of wastewater. The release of liquid and solid wastes in the stream caused discomfort to the population and degradation of the environment (Hydro-plante 2002). This also resulted in a considerable raise of the saline groundwater table thereby destroying the orchards. Farmers then began to use this water from the canal to irrigate their fields as a started practicing reuse by diverting the wastewater from the stream to their lands. Number of farmers installed cisterns and pumps for this purpose. Seemingly, wastewater was solely considered as a source of water; the fertilizing value was not recognized.

The irrigation of peach trees was started in 1995, about two years before the official creation of the wastewater irrigated area of Ouardanine. It was the initiative of a farmer from the region. An agricultural land stretching over 2 ha started to be cultivated with a new variety of peaches featured to give high yield and adapted to be grown intensively. The agricultural land was irrigated with the water from the Oued El Guelta, which was a mixture of wastewater effluents and stream water. As per the recollection of the farmers', less than two years later, a yield ranging 16–18 kg/tree was obtained. The fruits were of very high quality and sold in the local market at a very high price ranging 2–3.5 TND/kg (1.5 USD/kg).

Fig. 4 Location of Tunisia, Monastir governorate, and district of Ouardanine (Mahjoub et al. 2016)

4.1 Current Status

The positive economic outcome of above initiative overwhelmed farmers in the region and paved the way to the official creation of the wastewater irrigated area of Ouardanine. Figure 5 shows the evolution of the irrigated area and the volume of wastewater used for irrigation. After five years, the irrigated land has almost doubled in size with a subsequent increase in the wastewater use.

Based on the prevailing situation in Ouardanine in the early 1990s, the Ministry in charge of Agriculture and Water Resources has commissioned a study for planning the irrigation of 50 ha of agricultural land for a group of 36 farmers (CRDA 2015). The irrigation started effectively in 1997. Currently, the area has about 51 farmers and the irrigated area stretches over 62 ha. Crops irrigated with secondary effluents consist mainly fruit trees, comprising about 34 ha of peaches, pomegranates, figs, apples, and medlars. Forage crops like alfalfa and barley are grown as well on smaller area (CRDA 2015). It was noticed that the number of beneficiaries and the size of the actual irrigated area is approximate and variable according to the source of information due to lack of proper updates.

4.2 Management and the Quality Aspects of Wastewater

The Ouardanine WWTP was built in 1993 with a treatment capacity of 1500 m^3/d and a biological capacity of 600 kg BOD/d. It collected the effluents of 17,000 dwellers. In winter the maximum capacity may reach 1010 m^3/d. Wastewater is secondarily treated through a process of oxidation ditches (DGGREE 2015). Effluents are mainly domestic with few industries (slaughterhouse, perfume industry, olive mills, car washing station, etc.) that may impair the quality and cause troubleshooting to the treatment process (DGGREE 2015).

The irrigated area is jointly managed by GDA of Ouardanine and the CRDA who is in charge of the maintenance of the pumping station and other installed equipment. Farmers adhering to the GDA do pay a fixed annual fee of 15 TND

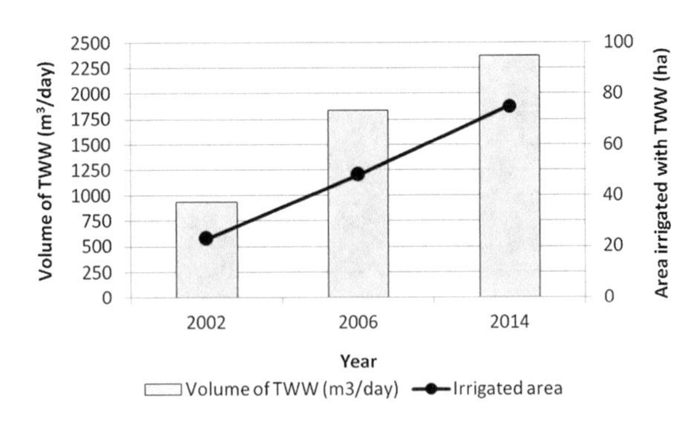

Fig. 5 Evolution of the volume of TWW (m^3/day) and the wastewater irrigated area (ha) in Ouardanine between 2002 and 2014. *Source* (CRDA Monastir 2016)

(6.5 USD) and a flat rate for wastewater supply based on irrigated land size (CRDA 2015; DGGREE 2015) since wastewater is available all year round and there are no functional meters to calculate the exact consumption. Water distribution is calculated based on the number of hours and the size of the irrigated area. Figure 6 shows the correlation between the size of the exploitation and the annual fees paid by the beneficiaries for the year 2015. Majority of them pay less than 400 TND/year reflecting on the small size of the lands, on one hand, and the low tariff of the reclaimed water, on the other hand.

5 Acceptance of Wastewater Use in Ouardanine: Data Collection

Since the time of its creation, Ouardanine has been considered the best wastewater irrigation case study that ever existed in Tunisia from technical and managerial points of views. Each year around 1000 visited the area to observe the process and learn the key reasons behind the success. However, there has been no studies conducted on the social acceptance and/or the motivations that lead to the success making this area well known throughout the world.

As the latest contribution, authors of this chapter conducted a survey on this topic in 2016. The survey included a questionnaire that was composed of 20 open-ended and closed questions organized in a semi-structured way. It was concise and focused on specific aspects identified beforehand, not to cause cognitive overload. The survey was carried out as a face to face interview; farmers did not have to fill in any forms. Based on the list of farmers provided by GDA Ouardanine, out of the 51 listed farmers, 13 (representing 25.4% of the community) have abandoned their land which is covering about 9 ha (15% of the total area). The remaining 53 ha are cultivated by 38 farmers; out of them 13 farmers are managing other irrigated land in addition to their own farms. This category of farmers was interviewed and only the responses related to their own lands were included in the survey. To collect objective opinions, farmers were questioned only about their own

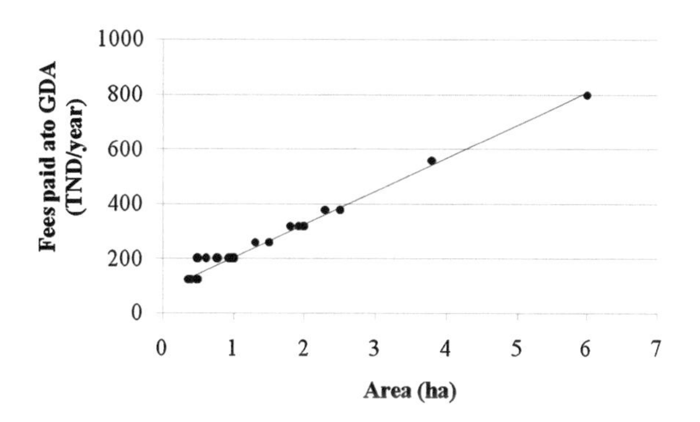

Fig. 6 Correlation between the fees paid to GDA (TND/year) and the size of the irrigated area (ha)

exploitation. The irrigated lands were split into 4 categories according to the size (Fig. 7). Based on the above statements and the categories described, the survey covered 18 farmers corresponding to 72% of the active farming community in the irrigated area. They are cultivating 32% of the wastewater irrigated area. Agriculture is a part-time activity for several beneficiaries that do not live on-site, therefore they were not available for the interview. This way of managing the land remotely is very significant with respect to the level of exposure to wastewater and health and environmental risks perception.

Farmers were interviewed on factors that may encourage or dissuade them from using TWW in irrigation. Indirectly, farmers have prioritized the main reasons behind the acceptance of reuse in Ouardanine. Information was collected in four categories: (i) the quality of TWW; (ii) commercialization of agricultural products; (iii) the volume of TWW available; and (iv) the regulations. About 60% of the interviewees think that the quality of TWW is the first factor that influence acceptance of use and pursuing an irrigation activity. Farmers fear the degradation of the quality through the discharge of industrial effluents in the sewer system because they think that this will have an impact on the irrigation system (e.g. possible clogging), human health (microbiological contamination), and the quality of the crops (contamination of the fruits). Users of TWW consider that consumers in the region of Ouardanine are not sufficiently sensitized to consumption of agricultural products irrigated with TWW. They estimate that consumers are still reluctant and have very low acceptance of reuse which may impact their income.

Another factor was the amount of water distributed for irrigation. Farmers may readily abandon practicing of TWW irrigation if supply becomes irregular or the amount of water is reduced in such a way that it does not meet the demand. Regulations were considered by them as an obstacle to acceptance. A more relaxed regulatory framework may encourage farmers to use TWW in agriculture and to engage in more profitable activities. Few farmers had mixed opinion, in general. These results were crosschecked through other questions about the benefits of TWW use which showed that 67% consider TWW as source of water and of fertilizers. High yield and good quality of the produced fruits are the second

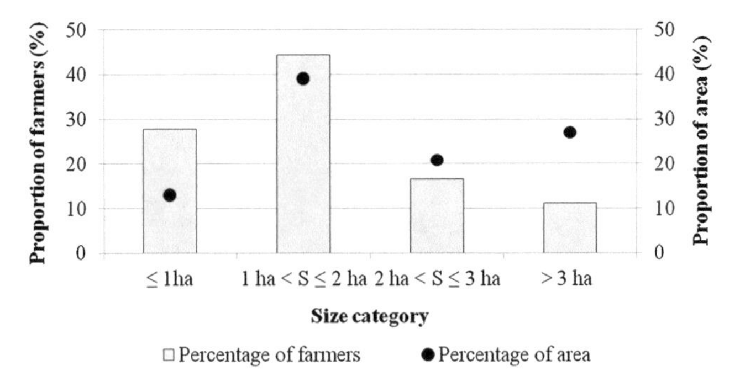

Fig. 7 Distribution of the irrigated land ownership according to size

advantage of reuse. The availability of TWW in terms of amount and timing of supply is regarded by 50% as one of the features.

6 Acceptance of Wastewater Use in Ouardanine: Results and Analysis

Based on the data collected through the survey conducted, the main factors influencing the acceptance of reuse of TWW in Ouardanine are discussed in detail in the following subsections.

6.1 *Financial Feasibility and Public Buy-in*

About 56% of the interviewees started reuse of wastewater during the creation of the irrigated area, after observing the results of the first pilot projects (Fig. 8). About 40% joined later. They either acquired new farm land leased from someone else.

Based on the present study, farmers' motivation to reuse was a combination of seven factors identified by the farmers themselves during the survey (Table 1). The responses were a mixture of one to four factors all together. About 44% of interviewees attributed their main motivation to practicing reuse of wastewater to the success of the first experience initiated in the area, in terms of financial benefit; it was the response with the highest frequency. Farmers recognized clearly the technical support and the guidance offered by the fellow farmer who started this practice first in the community. This supports the statements that financial feasibility, public buy-in and strong leadership are important factors to the success of reuse in such conditions.

In Ouardanine, years before the creation of the irrigated area, farmers used to grow garden crops in the stream bed of Oued El Guelta where the soil quality was recognized to be one of the best in the region owing to the deposition of sediments and organic matter. As a matter of fact, discharging wastewater into the stream was an obstacle to pursuing this activity, which resulted in was abandonment of it all

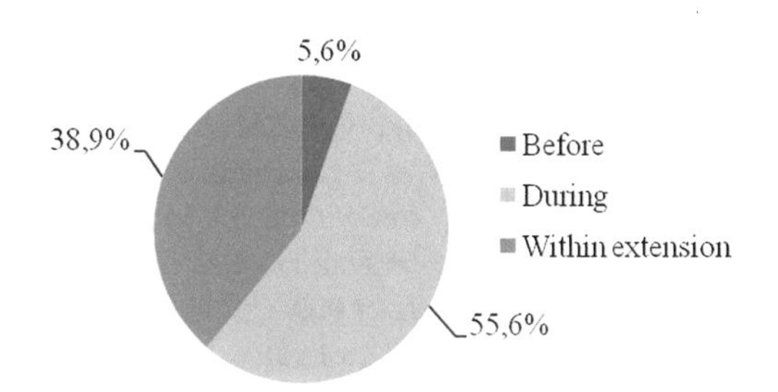

Fig. 8 Farmers' distribution with regard to the official creation of the project of reuse

Table 1 Identified factors behind the motivation to the reuse of wastewater in Ouardanine and their relative importance among interviewees

Identified factors	Frequency among farmers (%)
1. Financial success of the first experience and technical support	44
2. Absence of fresh water resources	33
3. Agriculture activity within the area (heritage, mutual support, yield)	33
4. Features of TWW (availability, quality, pricing)	17
5. Support of the CRDA and incentives related to the project	11
6. Reference to successful case study	6
7. Training on reuse of TWW in agriculture and economics	6

together. The economic loss caused by cessation of this activity had to be compensated, even though agriculture was not the main activity of the majority of the farmers. About 80% of them used to rely on other economic activities consisting mainly in leasing their farmland (60%) or having a part-time job out of the area. In fact, almost all the farmers used to live in the city, about 4 km away from the area (CNEA 2007).

As described above, in Ouardanine, the first initiative of reuse changed the livelihood. The economic well-being was the main driver to the acceptance of reusing this new water resource. To duplicate the experience and spread the success, the farmer that championed the project created a nursery to produce plants to sell in the surrounding area. The farmer stated that the use of wastewater has completely transformed his life from a small farmer to a large by giving him the financial affordability to invest in the agricultural and in other sectors. This requires a bunch of managerial qualities. Considering the outcomes of the first trial, the unexpected high yield and the high quality of the fruits were sufficient to inspire the neighborhood and to enhance the willingness to use wastewater for growing peaches.

The official creation of the wastewater irrigated area in Ouardanine was later extended. The study commissioned in 1997 only planned for a project of reuse covering 16 ha. Farmers who joined the project voluntarily had their agricultural lands located near the treatment plant. The trees cultivated were peaches, almonds, olive, and figs. Forage crops were also introduced for those raising livestock. The cultivated area quickly expanded to reach 30 ha after only three years. Another three years later, the irrigated area was covering 50 ha. About 25 ha of olive trees were planted as an extension of the project after the installation of a filtration unit and the adoption of water saving irrigation techniques.

According to farmers, the economic value of the agricultural land in Ouardanine has increased substantially. In the beginning of the 1990s, the value of one hectare of agricultural land was estimated at 5000 TND. During the project preparation the value of the land has increased to 20,000 TND and then to approximately 35,000 TND by the beginning of the project. Nowadays, and after supplying the region with potable water network and the rehabilitating the agricultural tracks, the value of one hectare of cultivated land is estimated at 100,000 TND.

In Tunisia, olive is a national agricultural heritage, especially in the central region Sahel where farmers generally tend to safeguard and may hardly approve removing their trees. However, the rural community in Ouardanine replaced the olive trees peach trees due to the economic benefits. Obviously, the benefit perceived in reusing wastewater was highly convincing and a driver toward a behavioral change. The economic feasibility was already established and proved therefore the launching of the feasibility study was considered just an official way to support to the project implementation.

Farmers were asked if they are willing to move back to conventional water resources instead of TWW if they become available. The large majority (78%) preferred to stick to TWW (Fig. 9). The main reasons behind were: (i) the availability of TWW throughout the year, (ii) the quality in terms of nutrients load, and (iii) the low price. Farmers consider relying on conventional water resources to be risky because of their possible low availability in terms of quantity and frequency of supply. They also admitted their satisfaction of the overall diversified nature of the agricultural development in the area. In contrast, the farmers who favored the use of fresh water were mainly concerned about the potential health issues related to wastewater irrigation.

6.2 *Awareness of Water Scarcity*

In Tunisia, use of TWW depends on the region and the location of the irrigated area, and usually correlates to the availability of fresh water; both surface and/or groundwater. The defining factor is the availability of water rather than the water quality. For instance, for the sake of saving their orchards farmers in Oued Souhil area had no other choice but to depend on TWW and groundwater, both found to be of low quality (El Amami et al. 2016). Naturally areas with availability and access to fresh water have developed reluctance and hesitation to adopt TWW irrigation (Abu Madi et al. 2003).

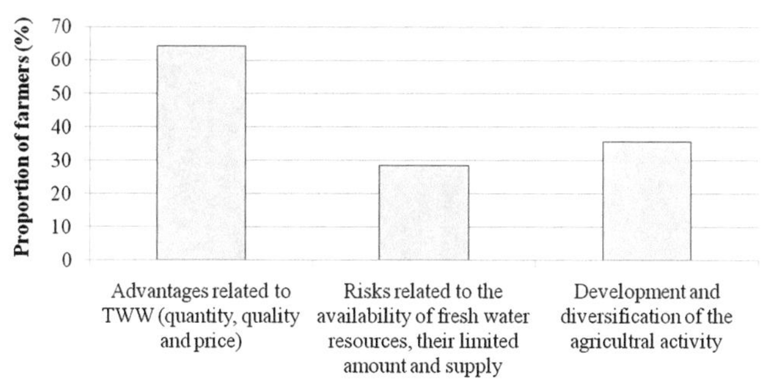

Fig. 9 Reasons behind keeping the activity of reuse even in case of availability of fresh water resources

The aquifer Sahline-Ouardadine which supplies water to the Ouardanine is highly saline (4.3 g/L) and overexploited (110%) (CNEA 2008). In 2003, groundwater quality monitored in 3 open wells (Fig. 10) registered 1.85–4.38 g/L salinity level in dry season and a concentration of nitrates ranging from 8.70–58.9 mg/L (BIRH-DGRE 2003). Besides, water of Nebhana dam, used exclusively for irrigation in Central part of Tunisia, is not available to the region of Ouardanine. Hence, wastewater is deemed to be the unique alternative to support agriculture in the area (Vally Puddu 2003).

Above findings from literature are in agreement with farmers' observations. In fact, 33% of the farmers interviewed acknowledged that the absence of alternative water resources was behind the decision to reuse reclaimed water for irrigation (Table 1). This factor is classified in second position. Indeed, farmers admitted being perfectly aware of water scarcity and the global phenomenon of climate change and the increasing trend of dry periods that may negatively affect the agricultural activities.

6.3 *Farmers' Commitment and Stakeholders' Involvement*

Involving farmers before and during the design of a project of reuse and its implementation is a prerequisite to it success. In Ouardanine, several farmers voluntarily joined the project after witnessing the success of the first experiment.

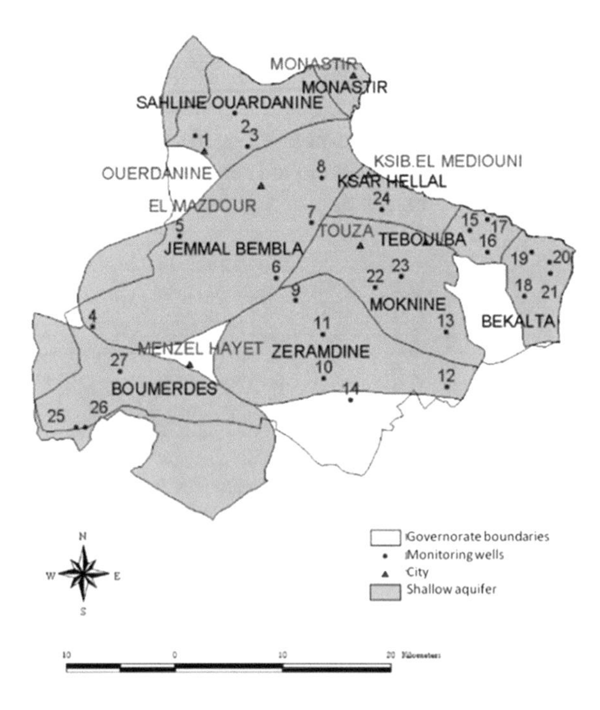

Fig. 10 Location of the aquifer Sahline-Ouardanine within Monastir Governorate and the monitoring wells (1, 2, 3) used for observing salinity and nitrates in Ouardanine (BIRH-DGRE 2003)

The survey showed that a little less than 50% of the farmers were involved in the design and preparation of the project; they were the ones who contributed to the project since its inception. The rest, almost 50%, have joined later during the project of extension of the irrigated area. These farmers were likely motivated by the success of their peers and did only have to copy on the progress made by their neighbors. Up-to-now, 89% of the interviewees declared being satisfied with the TWW use. However, 78% were willing to move to conventional water supply, if available.

6.4 *Knowledge and Education*

Numerous studies carried out worldwide have highlighted the importance of education in enhancing acceptance of wastewater in agriculture (Dolnicar et al. 2011; Hurlimann et al. 2008). Effectiveness of introducing basics, innovations, scientific outcomes and new concepts depends on the level of education of the audience.

As per the records of 2003, over 30% of the farmers between the 40–50 years were illiterate (Vally Puddu 2003). In the present survey, there were no illiterates; it was unintentional not to cover this category that may be the oldest in the area. Majority of the interviewees (33%) was older than 50 years and was from the first generation of farmers involved in the TWW use project since the launching. Generally, these farmers were of low educational level and have been only to primary school (Fig. 11).

Farmers in 20–40 age group represented 39% of the interviewee population (Fig. 12). This category also included university graduates (including engineers) who inherited farmland and did decide to maintain the agricultural activities on their lands. This category is a minority of highly educated beneficiaries. Successive heritage and dispersion of land ownership is considered an obstacle for agricultural activities in small lots, usually below 0.5 ha. This has resulted in land abandonment.

The age and the level of education influence the type of agricultural practices such as the estimation of water demand and amount of fertilizer to be added while taking into account the nutrients brought by wastewater during irrigation. Based on

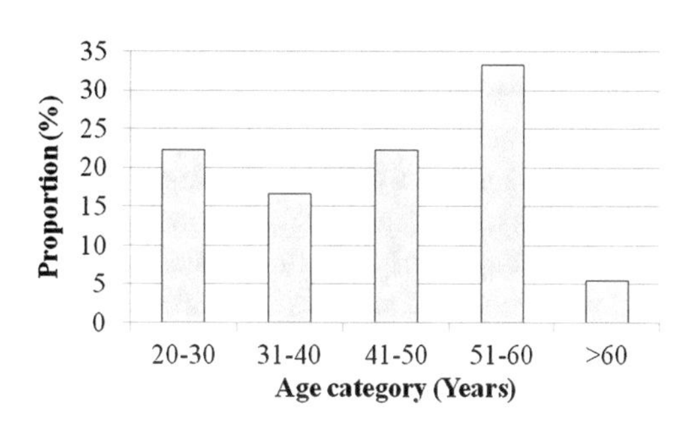

Fig. 11 Distribution of interviewees based on age categories

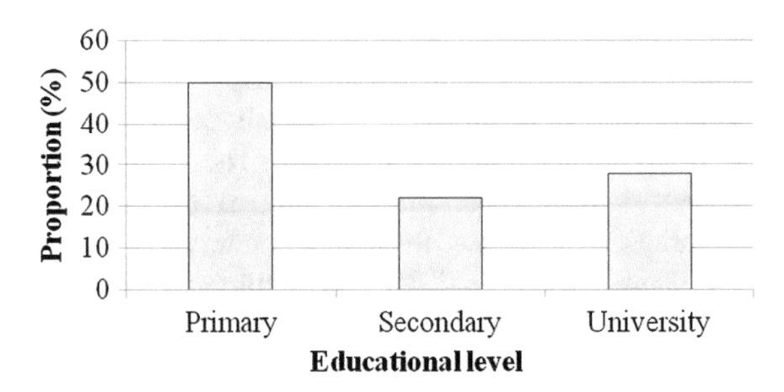

Fig. 12 Distribution of interviewees based on educational level

a study carried out in 2003, farmers admitted adding extra fertilizers because they could hardly estimate the quantities required for their trees or because they failed to consider that wastewater is an important source of nutrients (Vally Puddu 2003). The impacts of this practice are not yet evaluated on soil quality and on the sustainability of the agricultural system; therefore, they are worth a thorough investigation.

6.5 Fields of Reuse and Water Quality

To increase the acceptability of wastewater it is recommended to evaluate the quality of the TWW against the intended uses. In doing so, health and environmental risks can be mitigated more efficiently. Though, farmers may accept a lower quality of water if they assume that the consequences are under control (Drechsel et al. 2015).

Farmers' complains have been reported frequently in Ouardanine and was considered as an obstacle to the practice of reuse in the early stage of the project. Authors could not find any previous studies conducted on farmers' awareness on regulations and crops that they are permitted to grow with TWW. Use of wastewater for growing flowers in greenhouses indicates that the farmers do have some knowledge on the choices they have.

Farmers in Ouardanine are aware of the variable quality of wastewater. However, they do not have instruments to monitor water quality parameters to self-evaluate and decide if they are in compliance with the standards. If wastewater quality is not determined by the producers beforehand to approve its suitability for reuse, it will be evaluated subjectively downstream of the WWTP through its physical characteristics including color, odor, presence of foam, presence of visible pollutants, such as suspended matter. Otherwise, farmers can evaluate the quality of the water retrospectively through impacts on irrigation network (settling of sludge), irrigation system (clogging), soil properties (water logging), groundwater quality, etc. Clogging of drippers due to suspended solids occurred frequently in the past. Unfortunately, the solution had often been the removal of the nozzles or in the worst-case scenario abandoning drip irrigation all together. Nevertheless, farmers had no reluctance to use poor quality water, as the priority has been finding water somehow to irrigate at the right time. In fact, some farmers used a net/mesh to filter the load of visible suspended particles. All that flows downstream of the net was "supposed to be safe" to be used for irrigation (Mahjoub et al. 2016).

Based on analysis performed in 2015, TWW exceeded all threshold values stipulated in the national standards of reuse in agriculture: 182 mgO_2/L for BOD, 450 mgO_2/L for COD, and 117 mg/L for TSS versus the recommended values of 30, 90 and 30, respectively. The quality has substantially improved, since then. A sampling campaign carried out during November–December 2016 showed that COD, BOD and TSS are far below the standards. However, salinity is moderate with a mean of 1.55 mg/L at the WWTP outlet and high at the filters outlet with a mean of 3.10 mg/L, which is very high. This needs to be studied through a sampling campaign to identify the source of salinity. Such variation cannot be perceived instantly while it may have an impact on crops quality and yield. This issue is worth an extensive field study.

6.6 Communication

Farmers of Ouardanine were also asked about the existence of channels of communication with the technical services and the administration and the kind of support they have received to perform a safe reuse. About 17% stated receiving support from CRDA, GDA or CTV (local office of the extension service). Almost 90% did confess that they have never participated in any awareness campaigns. Majority of farmers received guidance from their neighbors, i.e. farmers who were already successful in implementing their projects, helped others. The first farmer who championed the practice of reuse in the region was considered a leader. Participation in trainings and getting in contact with the scientific communities and approaching international organizations did help him tremendously to achieve success. Based on the cumulated knowledge and experience of this farmer and his family, farmers of the region considered them as a substantial source of practical information.

It is important to highlight that Ouardanine is deemed as a very successful scheme of wastewater use in Tunisia. This has pushed some public authorities to see it as a sufficiently autonomous project that does not need any external guidance; which is not completely true.

In general farmers have a good relationship with GDA, when it comes to the wastewater distribution. However, when it is about the regional administration and local extension services, there is clearly a neutral relationship. It was also found that the farmers have a limited communication with CRDA (22%) and CTV (11%).

6.7 Exposure to Wastewater

In Ouardanine, majority of farmers do irrigate by themselves. Those living outside the region do irrigate during the week-ends and during vacation times. During the irrigation season, some farmers hire external labor. Participation of family members in agricultural activities, more specifically in irrigation, is very exceptional. This pattern indicates that the population most exposed to wastewater includes farmers and seasonal laborers.

The high acceptance of reuse in the region suggests that this population is aware of the inherent risks. Nevertheless, almost 90% of the interviewees did recognize not wearing protective tools during irrigation. This behavior may be stemming from the use of drip irrigation that it is thought to minimize exposure to pollutants, especially microbial pathogens, on one hand and the irrigated type of crops (fruit trees) that imply minimum contact with soil and irrigation water, on the other hand. Based on farmers' testimony, health risks are well managed within the irrigated area. This is supported by the absence of major incidents of related diseases such as diarrhea. As a matter of fact, and predictably, all farmers declared not taking vaccination.

Surprisingly, almost all farmers, except one with more knowledge on hepatitis and other water born disease, declared ignoring the usefulness of taking vaccination and the type of risks they are exposed to when they are reusing waste water. Prevention from infection by pathogens and chemicals and mitigation of health risks does not seem to represent an issue for the community. Even the first generation of farmers who took vaccinations at the project starting, did not follow afterward.

6.8 Health and Environmental Risks Perception

In Ouardanine, environmental impact made by wastewater discharge into the stream of Oued El Guelta was deleterious to the entire region. The aquifer near Oued El Guelta stream is only 4 m belowground. The discharge of the effluents in the stream resulted in: (i) growth of new and dense vegetation; (ii) progressive accumulation of sewage sludge; and (iii) rise of the water table in the surrounding area. The natural landscape was completely transformed into a saline ecosystem characterized by the invasion of halophyte vegetation stretching over an area as large as 5 km around the Oued El Guelta. The population noticed also the frequent occurrence of boars and other pests attracted by the new ecosystem, which were threatening the whole environment. It was reported that farm animals that used to graze and drink water from the stream died, giving proof to negative impact but also posing a great economic loss. These unhealthy circumstances drove dwellers, especially farmers, to accept and recognize the need for restoring, and preserving the natural ecological balance in the area.

Hence, diverting the effluents from the canal and use for irrigation was perceived as the best alternative to mitigate the burden of environmental damage caused by wastes dumping in the. The implementation of the project of reuse in the region of Ouardanine has contributed to the restoration of the natural system: the level of the water table has drop and pests and halophytes proliferation was reduced. A quantitative assessment including indicators is needed.

6.9 Attitudes of End-Users and Consumers

It is important to include not only the users of TWW but also the consumers of produce irrigated with TWW to evaluate public attitude towards TWW use. It is also possible to do it the other way around through collecting opinions of farmers on the marketability of their products and the channels they follow to sell their products.

Interviews conducted with farmers showed that consumption of the agricultural products irrigated with wastewater is not well perceived by consumers of the region. While use of TWW in agriculture has gained notoriety in a small city like Ouardanine, the rural communities started building a strong opinion about the topic and the quality of the agricultural products. Farmers declared that the wastewater irrigation has improved crop yield and the quality of the fruits, especially in peaches and pomegranates. Local and regional markets were the main channels for selling these fruits and consumers used to appreciate “this new product” known to be of very high quality. Private and public establishments like hotels were amongst the customers.

However, there had been other reports in the past about farmers encountering serious marketing issue caused by the poor acceptability. According to a survey carried out in 2003–2004, farmers were complaining about the significant decrease of peaches selling price. One unique way they have found to commercialize their agricultural products is by selling them in markets where there is no distinction between wastewater irrigated fruits and other types of fruits. Some also sell their fruits on-trees avoiding the cost of transportation to the local market thereby guaranteeing an acceptable income.

6.10 Gender Mainstreaming

The relationship between the discomfort and the individual differences has shown that women, compared to men, feel more uncomfortable to use wastewater because of the disgust and sensitivity to pathogens (Wester et al. 2015). Thus far, no other study has shown similar attitudes of women in Tunisia in this regard except for scarce observations. The role of Tunisian women in wastewater irrigation is not well documented. There is no official data estimating or quantifying the role of

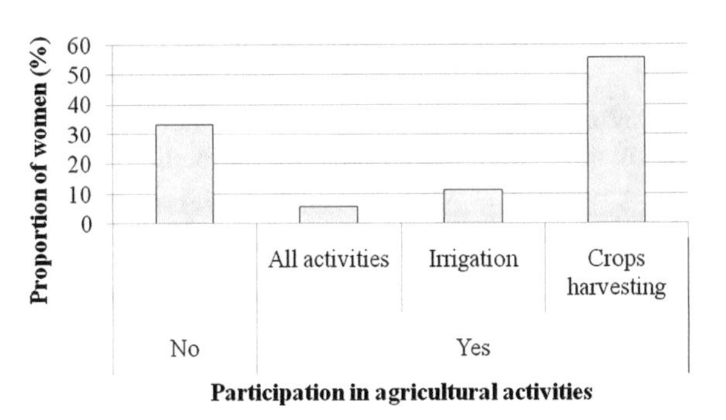

Fig. 13 Distribution of women's role in the activity or reuse

women working in use of wastewater, permanently or occasionally, despite their important role in preserving health either within the family or within their environment.

In Ouardanine, nearly 10% of farmers are women owning 13% of the total irrigated land. These women are managers but they do not necessarily take decisions. There was only one woman managing her own land. Many women are usually employed as seasonal workers. The survey showed that about 70% of the interviewees declared involving women in the agricultural activities related to the use of wastewater. Women are mainly participating in fruits and crops harvesting (Fig. 13). They are mostly hired as hand labor while a few is engaged in irrigation and other activities in the field. Wives of farmers may also help in field work but this was not quantified. The occasional presence of women in the field implies their exposure to wastewater through moist soil after irrigation.

7 Conclusion

Use of wastewater can reduce vulnerability of population and agricultural activity to drought and provide number benefits to dwellers. Public acceptance influences success or failure of such projects. Positive social and cultural acceptance can act as a precursor of the success if handled appropriately as it is associated to a myriad of interlinked factors such as knowledge, financial feasibility, environmental and health risks perception, etc. Tunisia was among the first few countries that realized the potential of TWW in agriculture. However, comprehensive studies on public acceptance are still rare in Tunisia. Achieving the targeted of using 50% of its wastewater by 2020, Tunisia requires an establishment to adequately address the reluctance that some of the farmers and consumers have shown.

The irrigated area of Ouardanine is an interesting case study where farmers succeeded in practicing wastewater irrigation and achieve economic benefits without compromising the safety. A thorough analysis of the factors behind the acceptance of wastewater use in the region of Ouardanine enabled ranking the main drivers and by the same token highlighted challenges and opportunities for future improvements. In brief, information collected through a survey allowed identifying seven key factors as promoters of the acceptance of wastewater use in the region. The financial benefit was ranked at the top. It proved that adopting circular economy concept is very crucial through creation of value and gain of tangible benefits. Another key finding is the factor ranked as the second: the absence of conventional water supply as alternative water resources for irrigation. Any change in wastewater quality and/or its supply may compromise the practice of reuse in the region.

The current relationship between farmers and the regional/local stakeholders is weak; to continue the trend of success this needs to be improved through a better communication plan. The presence of an unofficial "leader" in the community was seen positively, as he could minimize the gap of knowledge among farmers. However, this is a case specific situation that may not happen in other regions of the country. Attitude of consumers is largely overlooked and public distrust is a barrier to commercializing wastewater irrigated products in the region. Guaranteeing outlets would encourage farmers to engage in wastewater use projects and sensitization of the public would reduce its reluctance. Gender issues seemed to play a minor role in the area. These findings that specifically come from Ouardanine could support future initiatives elsewhere, within Tunisia or even in other countries. It is important to address perception of reuse in other irrigated schemes to ameliorate public acceptance.

Acknowledgements Authors would like to acknowledge the support received from the farmers, especially Mr Mohamed Mekada, and also CTV Ouardanine, CRDA Monastir, and GDA Ouardanine during the data collection. All helped to conduct the survey in a short time without compromising the quality. Special thanks to GDA President for his help in approaching farmers and making them available to assist.

References

Abu Madi, M., Braadbaart, O., Al-Sa'ed, R., & Alaerts, G. (2003). Willingness of farmers to pay for reclaimed wastewater in Jordan and Tunisia. *Water Science and Technology: Water Supply, 3*, 115–122.

AVFA. (2008). http://www.avfa.agrinet.tn/ Accessed February 15, 2017.

Bahri, A. (1998). Fertilizing value and polluting load of reclaimed water in Tunisia. *Water Research, 32,* 3484–4389.

Baumann, D. D. (1983). Social acceptance of water reuse. *Applied Geography, 3,* 79–84.

BIRH-DGRE. (2003). Directory of groundwater quality in Tunisia (In French).

Buyukkamacia, N., & Alkan, H. S. (2013). Public acceptance potential for reuse applications in Turkey. *Resources, Conservation and Recycling, 80,* 32–35.

CNEA. (2007). Evaluation of the current situation in nine schemes irrigated with treated wastewater. The irrigated area of Ouardanine, Monastir governorate. Phase 2: Diagnosis of the current situation and recommendations (In French). 51 pages + annex.

CNEA. (2008). Study on the desertification for a sustainable management of natural resources in Tunisia (in French). Report on the third phase. Last Accessed: February 2008, http://www.chm-biodiv.nat.tn/sites/default/files/rapport_desertif.pdf.

CRDA. (2015). Experience of the GDA of Ouardanine 2 in the reuse of treated wastewater (in Arabic).
DGEQV. (2013). Stratégie nationale de communication et de sensibilisation à l'utilisation des eaux usées traitées et des boues de STEP et initiation des activités de sensibilisation à l'échelle regionale. Phase 2: Elaboration de la stratégie. Ministère de l'Equipement et de l'Aménagement du Territoire et du Développement Durable. 102 pages.
DGGREE. (2015). Report on the situation of the areas irrigated with treated wastewater Campaign 2014/2015. (In French).
DGGREE. (2016). Report on the current status of the reuse of treated wastewater in irrigated areas. 46 pages. (Unpublished internal report).
Dolnicar, S., Hurlimann, A., & Grün, B. (2011). What affects public acceptance of recycled and desalinated water? *Water Research, 45,* 933–943.
Drechsel, P., Mahjoub, O., Keraita, B. (2015). Social and cultural dimensions in wastewater use. In P. Drechsel, M. Qadir, & D. Wichelns (Eds.), Wastewater: An economic asset in an urbanizing world. Geographical focus: Low- and middle-income countries. Part 3. The enabling environment for use, Springer, 75–92. ISBN:13 978-9401795449.
El Amami, H., Mahjoub, O., Zaïri, A., Mekki, I., & Bahri, H. (2016). Conjunctive use of recycled water and groundwater: An economic profitability and environmental sustainability analysis - Oued Souhil case Study. In *Proceedings of the 2nd International Conference on Integrated Environmental Management for Sustainable Development*, Volume 2: Water resources, 27–30 October 2016. Sousse, Tunisia: Springer. ISSN 1737-3638
Grundman, P., & Maas, O. (2017). Wastewater use to cope with water and nutrient scarcity in agriculture- A case study for Braunschweig in Germany. In J. R. Ziolkowska, & M. Peterson, J. M.(Eds.), *Competition for water resources. Experiences and management approaches in the US and Europe* (pp. 352–365). ISBN: 978-0-12-803237-4.
Hartley, T. W. (2006). Public perception and participation in water reuse. *Desalination, 187,* 115–126.
Hurlimann, A. C., Hemphill, E., McKay, J., & Geursen, G. (2008). Establishing components of community satisfaction with recycled water use through a structural equation model. *Journal of Environmental Management, 88,* 1221–1232.
Hydro-plante. (2002). Etude d'Assainissement et de Recalibrage de l'Oued El Guelta (in French). Dossier d'éxecution, Tunisia.
INFAD. (2012). Wastewater Treatment. World Fatwa Management and Research Institute, Islamic Science University of Malaysia. Retrieved on March 23, 2012 from http://infad.usim.edu.my/.
INNORPI. (1989). Use of reclaimed water for agricultural purposes - Physical, chemical and biological specifications (in French). *Tunisian standards NT, 106*(03), 1989.
ITES. (2014). Strategic study: Hydraulic system of Tunisia at the horizon 2030. (In French) Tunisia.
Mahjoub, O., Mekada, M., & Gharbi, N. (2016). Good Irrigation Practices in the Wastewater Irrigated Area of Ouardanine, Tunisia. In H. Hettiarachchi & R. Ardakanian (Eds.), *Safe use of wastewater in agriculture: Good practice examples.* UNU-FLORES, 101–120, ISBN: 978-3-944863-31-3.
Neubert, S., & Benabdallah, S. (2003). Reuse of treated wastewater in Tunisia. Studies and expertise reports. (In French). 11/2003. Bonn.
ONAS. (2015). Annual report. 27 pages.
Özerol, G., & Günther, D . (2005). The role of socio-economic indicators for the assessment of wastewater reuse in the Mediterranean region. In A. Hamdy, F. El Gamal, N. Lamaddalena, C. Bogliotti, & R. Guelloubi (Eds.), *Nonconventional water use: WASAMED project. Bari: CIHEAM/EU DG Research,* (pp. 169–178). Options Méditerranéennes: Série B. Etudes et Recherches, n. 53.
Republic of Tunisia. (2016). National Development Plan. Volume I. 189 pages.
Rejeb, S. (1990). Effects of wastewater and sewage sludge on the growth and chemical composition of some crop species (in French), Tunisia, 164 pages.

Selmi, S., Elloumi, M., & Hammami, M. (2007). La réutilisation des eaux usées traitées en agriculture dans la délégation de Morneg, en Tunisie. Mohamed Salah Bachta (Ed). Les instruments économiques et la modernisation des périmètres irrigués, 2005, Kairouan, Tunisie. Cirad, 10 pages.

Trad-Rais, M. (1988). Microbiologie des eaux usées traitées et quelques résultats de leur valorisation fourragère. Tunisie, CRGR, 9 pages.

Vally Puddu, M. (2003). *Technico-economic diagnosis of wastewater use in the irrigated area of Ouardanine (Monastir-Tunisia) (in French)*. INRGRF: Graduation project. Tunisia.

Wester, J., Timpano, K. R., Cek, D., Lieberman, D., Fieldstone, S. C., & Broad, K. (2015). Psychological and social factors associated with wastewater use emotional discomfort. *Journal of Environmental Psychology, 42,* 16–23.

World Health Organization (WHO). (1989). *Health Guidelines for the Use of Wastewater in Agriculture and Aquaculture*. Geneva, Switzerland: Report of WHO Scientific.

World Health Organization (WHO). (2006). WHO guidelines for the safe use of wastewater, excreta and greywater. Volume II: Wastewater in Agriculture. Geneva, Switzerland: WHO-UNEP-FAO.

World Health Organization (WHO). (2009). Religious and cultural aspects of hand hygiene. WHO guidelines on hand hygiene in health care: First global patient safety challenge clean care is safer care. First global patient safety challenge clean care is safer care. Geneva: World Health Organization. ISBN-13: 978-92-4-159790-6.

Zekri, S., Ghezal, L., Aloui, T., & Djebbi, K. (1997). Negative externalities of the reuse of treated in agriculture (In French). Options Méditerranéennes, SérieA, No31, Séminaires Méditerranéens.

Use of Wastewater in Managed Aquifer Recharge for Agricultural and Drinking Purposes: The Dutch Experience

Koen Zuurbier, Patrick Smeets, Kees Roest and Wim van Vierssen

Abstract Use of wastewater is increasingly gaining importance as a water supply. However, the acceptance of the final users is important for the success of such projects. The acceptability of the treated wastewater depends on the physical, chemical, and most importantly the microbiological quality of the water. Appropriately designed and operated Managed Aquifer Recharge (MAR) systems have proven to be a very effective and robust barrier against all pathogens present in wastewater. Examples of successful implementation of MAR to catalyse safe and reliable water reuse are abundant. In the Netherlands, this started with the intake river water for dune infiltration in the 1950s. These big MAR schemes still supply around one-fifth of the drinking water in the Netherlands. Research has shown that these MAR systems are crucial for disinfection of the river water and overcoming mismatches between river water availability and water demand. Cost-effective and microbiologically reliable water supply can also be attained for the agricultural sector, as shown by the Dinteloord case study. Stakeholder involvement and an integrated approach is becoming indispensable for MAR and results in increased creation of water banks, including total cost recovery based on financing from all stakeholders.

Keywords Wastewater · Managed aquifer recharge (MAR)
Aquifer storage and recovery (ASR) · Water quality indicators
Dune infiltration · Pathogens · Disinfection

K. Zuurbier (✉) · P. Smeets · K. Roest · W. van Vierssen
KWR Watercycle Research Institute, P.O. Box 1072, 3430 BB Nieuwegein, The Netherlands
e-mail: Koen.Zuurbier@kwrwater.nl

W. van Vierssen
e-mail: Wim.van.Vierssen@kwrwater.nl

P. Smeets
Utrecht University, P.O. Box 80125, 3508 TC Utrecht, The Netherlands

W. van Vierssen
TU Delft, P.O. Box 5, 2600 AA Delft, The Netherlands

H. Hettiarachchi and R. Ardakanian (eds.), *Safe Use of Wastewater in Agriculture*,
https://doi.org/10.1007/978-3-319-74268-7_8

1 Introduction

The increasing mismatch in freshwater demand and availability in many part of the world has resulted in many innovative means to exploit the limited available water sources such as rainwater, surface water, groundwater, seawater and wastewater. The challenge is to provide sufficient water of sufficient quality for the intended use. Water scarcity is driving force behind the utilization of various unconventional water sources that may even contain chemical and microbiological contaminants and threaten public health. New, sustainable water supply concepts should address not only the steady supply of it, but also the safety.

Managed aquifer recharge (MAR) is well-recognized as a leading-edge strategy to improve water quality and provide storage in a combined nature-based solution (Dillon 2005; Dillon et al. 2006, 2010; Pyne 2005), in which aquifers can provide an ecosystem service (DESSIN 2014). MAR includes various ways of artificially recharging aquifers such as surface spreading (basins, ditch and furrow, flooding, soil aquifer treatment, percolation tanks), injections wells, sprinkling, and groundwater conservations structures (e.g. subsurface dams). MAR is used for various aims, the most important ones being storage (protected against algal blooms, fallout, and evaporation), purification (removal of pathogens and micropollutants without need of disinfection or oxidation), attenuation of quality fluctuations (including temperature), maintenance of groundwater levels (to prevent mining, seawater intrusion or preserve wetlands), and transportation (aquifer as a conduit).

Interesting examples of MAR applications to support reliable water supply can be found in Australia, the USA and Europe. In the Netherlands, MAR has been applied on a small scale since 1940. In the 1950s large projects were established to supply water to the western coastal zone of the country. In this densely populated area, groundwater abstraction is restricted due to salt water intrusion and decline of groundwater levels. Yearly, 177 Mm^3 of surface water is infiltrated for drinking water production in the Netherlands (16% of total production). The most important reason to apply MAR in the Netherlands (Dutch setting) is to improve the quality of the water from the River Rhine (Fig. 1) and Meuse which has a poor quality as a consequence of industrial/municipal wastewater discharges and agricultural/urban drainage in upstream. Additionally, the MAR systems can be used during the several months the river water quality is very poor.

Enabling a reliable and safe water supply for agriculture and drinking water requires a careful design and integral evaluation of pre-treatment, aquifer processes (during MAR), post-treatment, and distribution. In this chapter, we present recent insights in wastewater reuse and the use of MAR to catalyse this, which are based on decades of MAR development and recent innovations in the Netherlands and the European Union.

Fig. 1 Primary source of the Dutch drinking water: The River Rhine. The spring of the river is in the Alps, while its course passes various big industrial areas and cities. Illustration by: Paul Maas Illustratie

2 The Water Supply in The Netherlands: The Dutch Secret

In the 19th century, groundwater wells and surface water were the main sources of drinking water in the Netherlands. In large cities, the canals used for water supply became contaminated by the increase of population and groundwater was insufficient as a resource. Large outbreaks of typhoid and cholera were common. In 1853 fresh water from the dune area was transported through pipes into the city of Amsterdam as an attractive and safe alternative, even though the knowledge of waterborne diseases was not common at that time. Initially water was supplied through standpipes, but gradually house connections were made. Other large cities near the coast followed and large dune areas became protected sources of water supply. The popularity of dune-water meant the fresh water reserves in the dunes became depleted, attracting brackish water. Since 1950 water from the river Rhine was pretreated, transported to the dunes and infiltrated through canals to replenish the fresh water reserves.

2.1 Use of MAR for Reuse of Surface Water Fed by Treated Wastewater

Large MAR facilities are now in the coastal dune areas along the North Sea (Fig. 2). These areas are characterized by permeable, sandy, and unconfined

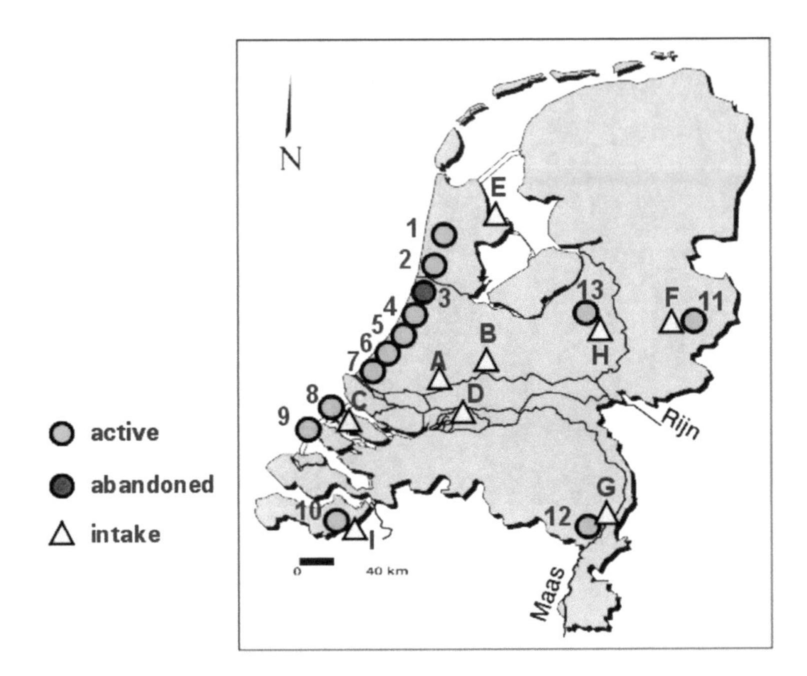

Fig. 2 MAR sites using basin infiltration of pre-treated surface water and location of the river water intake (Stuyfzand and Doomen 2005)

aquifers. Basins are therefore used to infiltrate the surface water, primarily from the Lake IJssel (17%), the River Rhine (35%), and the River Meuse (39%). Recovery occurs by open basins and extraction wells. In 2014, approximately 177 Mm^3 was artificially infiltrated this way, over an open area of 307 ha (Vewin 2014). This amounts to almost 16% of the drinking water produced in The Netherlands. The yearly volume of drinking water produced by MAR has been relatively stable during the last decades. Besides use for drinking, this tap water is also used by industries and high-end agriculture (mainly greenhouses).

Before infiltration, the water is generally pre-treated with coagulation and flocculation, flotation or sedimentation, and rapid sand filtration. The transport to the infiltration sites is 60 km on average (Peters 1995). The residence time in the aquifer are in the range of 20–200 days. The open infiltration proofs to be valuable technique with which unreliable water can be turned into a hygienically safe source for drinking water productions. The main reasons to go for this technique in the coastal zone were:

- Continuity of drinking water supply;
- Prevention of saltwater intrusion after extensive groundwater extraction;
- Purification of the surface water during infiltration;
- A constant quality by mixing in the aquifer during transport; and
- Mitigation of falling groundwater levels.

The redox environment proved to be the chemical master-variable for these MAR systems in The Netherlands, controlling to a high degree the mobility, dissolution, breakdown, and toxicity of many inorganic and especially organic compounds in or in contact with the water phase (Stuyfzand and Doomen 2005). For many micropollutants, the specific 'redox barrier' is established for degradation or precipitation.

To reduce the impact on landscapes and ecology in the dune area, deep well injection schemes were developed for direct injection into deeper aquifer. With this technique, rapid clogging of the recharge wells can be avoided by proper aquifer exploration, optimal well design, construction and operation, and early regeneration (Olsthoorn 1982; Peters 1995). The largest deep well injection site is Watervlak, consisting of 20 infiltration wells with a capacity of 5.3 Mm^3/yr.

2.2 Ensuring the Microbiological Reliability

When MAR was introduced, most cities already had municipal sewerage and all this sewage eventually ended up in the rivers with limited or no treatment. The water supply relied on the infiltration in the dunes and chlorination to avoid spreading diseases through drinking water. When it was found that the disinfection by-product of chlorination could cause cancer in 1974 (Rook 1976), a gradual ban was introduced against the use of chlorine as a main disinfection barrier. This required extra care for the microbial safety of the water. Although there were no water-supply related outbreaks in the Netherlands, outbreaks in other countries made clear that presence of chlorine and absence of *E. coli* bacteria alone were insufficient to guarantee drinking water safety. The dune-filtration stage was considered an important barrier against pathogenic micro-organisms that were present in the river water in high numbers. This did raise the question whether this barrier was effective enough, and how many additional barriers were needed to achieve safe drinking water. In 2001 a legal health target was set at 1 infection per 10,000 people per year, to be verified by regular Quantitative Microbial Risk Assessments (QMRA) of the surface water supply systems (Anonymous 2001). Rather than using general 'log credits' as applied in other regulations (Australian government 2008) the site specific assessment requires sufficient monitoring of the actual site (Wetsteyn 2005).

This meant that the microbial safety of the MAR systems in the dunes needed to be assessed from the river source to the produced drinking water. Pathogens were monitored directly in the river water or after storage in intake reservoirs. Even though most of the discharged wastewater is now treated, the river water contains pathogenic viruses, bacteria and protozoa. Concentration of them in the river waters is only two orders of magnitude lower than in the untreated sewage, because tertiary wastewater treatment and dilution only have a limited effect on pathogens. Wildlife, agricultural runoff and combined sewer overflows can also contribute to pathogen loading in surface water, especially *Campylobacter* from waterfowl is a zoonotic

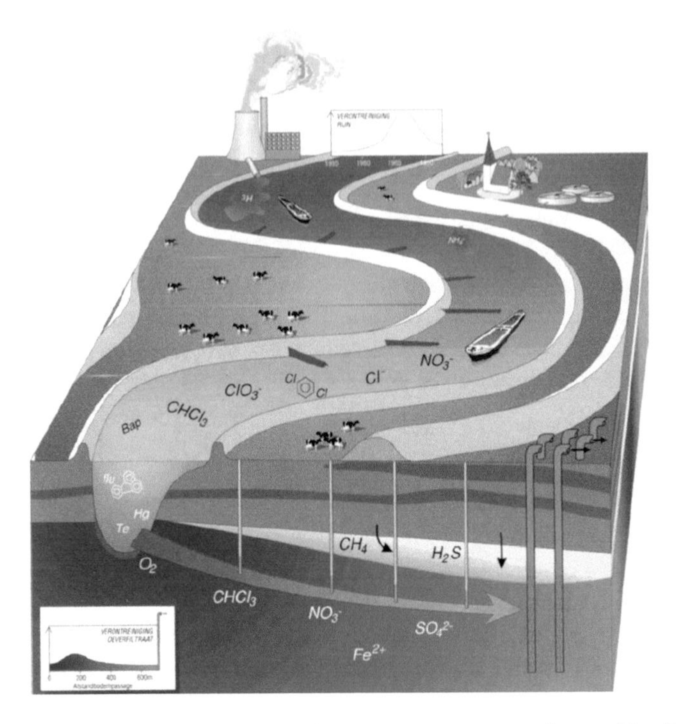

Fig. 3 Schematization of river bank filtration at water company Oasen, The Netherlands (Stuyfzand and Doomen 2005)

pathogen of concern. In order to achieve the health target of another 4–8 log reductions of pathogens is required, depending on the type of pathogen and location (Smeets et al. 2009). In the MAR systems this is achieved by multiple treatment barriers consisting of pretreatment, MAR and post treatment. Pretreatment consists of conventional treatment (coagulation-sedimentation and rapid sand filtration) after which the water is transported to the infiltration ponds. The MAR systems are designed to achieve a minimum of thirty days residence time. Field experiments have shown that this is sufficient to achieve a removal of 9 log cycles (Schijven et al. 1998; Van der Wielen et al. 2008; Hornstra et al. 2013), making this the most effective barrier.

Several researchers have modeled the attachment of pathogens to soil particles and their die-off during soil passage. The model by Tufenkji et al. (2004) is currently widely applied. For typical MAR conditions in the Netherlands, this is the most important process of removal, and inactivation is limited when residence time is short. Infiltration and abstraction from the same well would require post disinfection because the last water in is the first water out, with short residence time and filtration distance. Therefore, MAR infiltration and abstraction points in the Netherlands are set some distance apart, providing a minimum filtration distance and travel time resulting in verified pathogen removal.

Post treatment focuses on parameters such as hardness, biological stability, taste and odor. The processes used may include ozonation, granular (biological) activated carbon filtration, pellet softening and slow sand filtration. Especially ozonation and slow sand filtration processes form additional barriers against pathogens, even though that is not their main goal.

Some 10 bank filtration systems (Fig. 3) abstract water from the banks of rivers or lakes, thus implementing a basic form of MAR. These are planned such that the minimal residence time of a single stream line is at least 30 days, thus achieving over 9 log removals of viruses due to dilution with longer streamlines and other wells in the well-field.

Especially during dry summer period, the proportion of wastewater in the rivers can exceed 10%, meaning these systems can be regarded as indirect reuse of wastewater. One Belgian system uses treated wastewater directly as a source for MAR. In this case the pretreatment is more extensive, including reverse osmosis membrane filtration as an extra barrier against both microbial and chemical contaminants.

3 Wastewater in MAR: The Case of Dinteloord

In the western part of the Province of Noord-Brabant, near the shoreline of the Dutch southwestern estuary, a 220 ha high-tech greenhouse cluster is under development (Fig. 4). With fresh groundwater being a scare resource in the generally brackish groundwater system and river discharges being already critically low, a reliable and sustainable freshwater supply for the typical months of demand (April–August) was a big challenge for the development of this greenhouse cluster. A solution was provided by the nearby sugar factory, which produces sugar from sugar beets in the harvesting season (September until January, each year). The wastewater (more than 1 Mm^3/yr) coming from this sugar factory is extensively treated before disposal on the River Dintel. By adding a polishing step and a combination of submerged ultrafiltration (UF) and reverse osmosis (RO), a part of the effluent (up to 0.3 Mm^3/yr) is upgraded to high quality, virtually demineralized, irrigation water (EC: 0.01 ms/cm).

To overcome the mismatch between the production of the high-quality irrigation water and the demand a period of 2–6 months needs to be bridged by storage. The first plans were to store the water in aboveground open basins, constructed by 4 m high artificial dikes covered with EPDM foil. However, the developers of the area had serious doubts on the conservation of the water quality during the long-term storage in this costly (approximately 2.5 million $) reservoir, which would be accessible for birds and unauthorized persons, while being unprotected from atmospheric deposition and evaporation. MAR provided a solution in the form of ASR, which could easily be implemented on unexploited land surfaces in an ecological zone in the heart of the area.

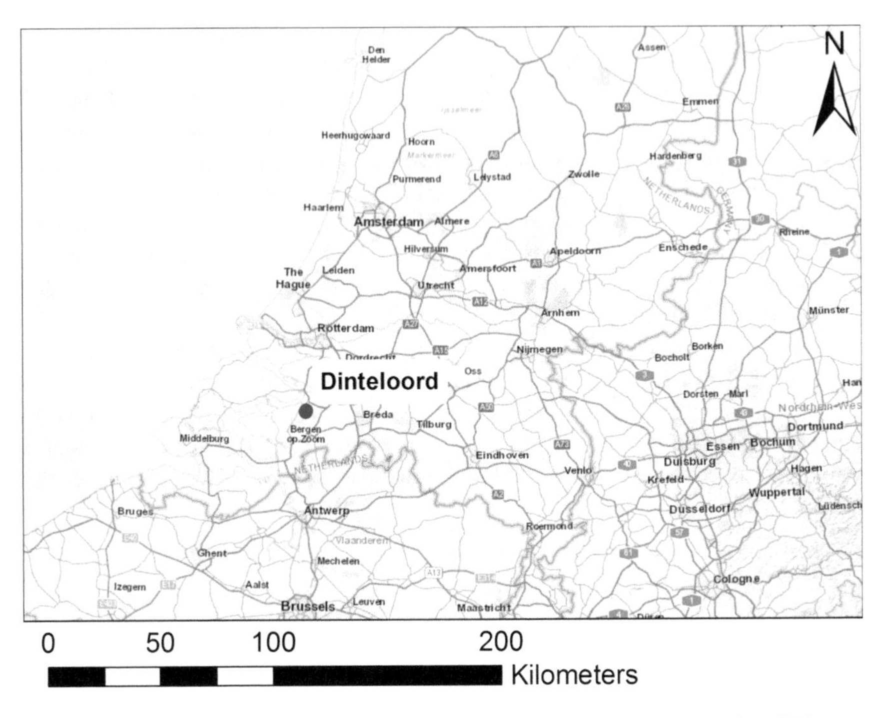

Fig. 4 Location of the Dinteloord project area. (Map source: World Street Map by ESRI)

By using ASR in a greenhouse set-up (Zuurbier et al. 2014, 2017; Zuurbier and Stuyfzand 2017), the high-quality water is protected from the external impacts threatening the water in the open reservoir. In total, eight wells are required to supply at least 200 m^3/h to guarantee a supply of 1 m^3/h per ha of greenhouse in the area, which is part of a service pack for growers buying land in the area. The wells are installed in a fine sand aquifer at a depth of 10–30 m below surface level, which is covered by clay and peat deposits. The ambient groundwater is considered slightly brackish for greenhouse irrigation (sodium: 40 mg/l, while 2.3 mg/l is demanded), while also Cl, Ca, HCO_3, Fe, Mn, and various other elements exceed the limits of irrigation water. Figure 5 provides a schematic of the solution.

The realisation of the ASR scheme in Dinteloord took around 5 years to carefully assess the feasibility, select appropriate locations, obtaining the required permits, and to validate the perspectives in a small-scale field pilot, before upscaling (Table 1). In this pilot (Fig. 6), an extensive monitoring of the water quality development took place, to validate the development of the chemical and microbiological quality of the water upon recovery and the hydrological effects on the surroundings. This was not only required for operational optimization, but also to gain confidence from stakeholders (neighbouring farmers, municipalities, the sugar factory, etc.) and was organized in the Subsol EU-project (Grant Agreement No 642228).

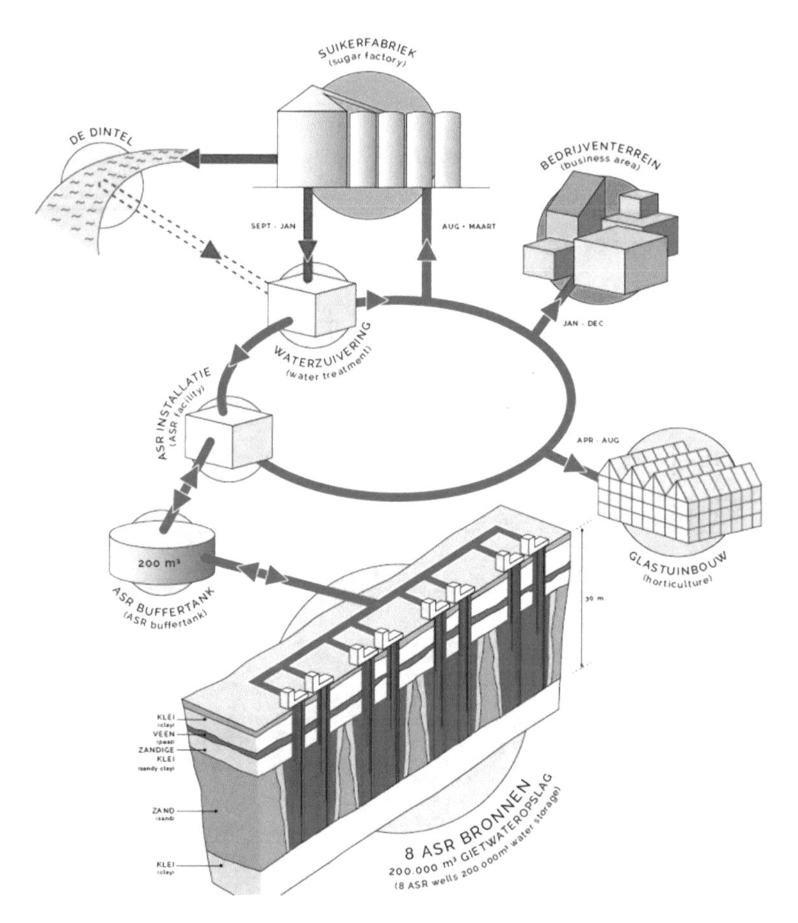

Fig. 5 The Dinteloord water system, using effluent from a sugar factory for irrigation water supply upon aquifer storage and recovery

The first ASR-cycle was simulated using SEAWAT, a groundwater transport model (Fig. 7) in order to improve the insight of freshwater distribution in the target aquifer and assess the potential recovery. It was found that the infiltrated freshwater body remained at a stable position near the ASR well, which was underlined during recovery in 2017, when all of the infiltrated water was successfully recovered with minimal admixing of ambient brackish groundwater. Based on water analyses, the water quality was found affected by calcite dissolution and pyrite oxidation, which led to a slight but acceptable increase in Ca, Mg, HCO_3, SO_4, Fe, and Mn. Harmful viruses and bacteria were not observed during analyses.

When upscaling of the well field was simulated, it was found that a virtually complete recovery of freshwater within the quality limits set is feasible, despite some increasing concentration in the recovered water at the end of the recovery

Table 1 Road to implementation of ASR in Dinteloord

Phase	Time	Duration	Activities	Products
1	2012	2 months	Desk-study	Feasibility assessment
2	2012	2 months	Sampling existing wells	Improved feasibility assessment
3	2013	3 months	Pilot drilling	Preliminary design
4	2013	2 months	Environmental impact assessment	Report
5	2014	6 months	Permitting	Permit
6	2015	4 months	Drilling, installation	First ASR-well
7	2016	8 months	Pilot ASR cycle	Field measurements, calibrated groundwater model, evaluation, final design
8	2016	1 day	Stakeholder meeting	Engagement of water users and related actors
9	2017	8 months	Upscaling	ASR-well #2–4
10	2017–2018	24 years	Monitoring, evaluation	Final evaluation of ASR performance before final upscaling (well #5–8)

Fig. 6 The first ASR-well (PP1) in the well field in Dinteloord (left) and its well completion including control valves in the well house (right)

stage, mainly in the first cycles (Fig. 8). The first species attaining a critical concentration is Na, which limit is set to only 2.4 mg/l.

4 The Economic Aspects of Wastewater in MAR

For the developer and the growers in the Dinteloord area, going for MAR was not only based on the better protection of the water quality. Due to the high costs of land (otherwise used for high-end agriculture) and construction of the 300,000 m^3 reservoir, the price of storing the water upon treatment would be more than Euro 3.00/m^3 of water supplied (Table 2), mainly due to frequent re-investments in the

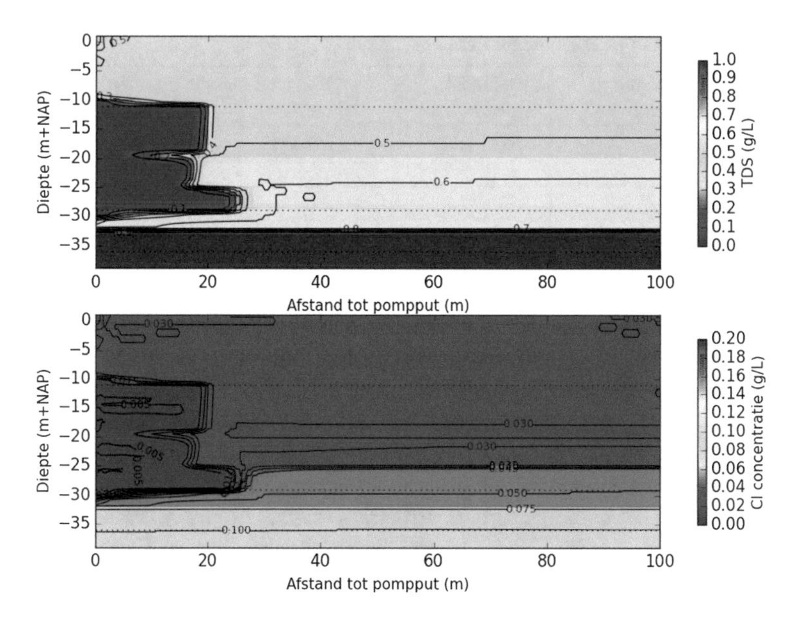

Fig. 7 TDS and Cl concentrations in the aquifer during storage (June 3, 2016). Depth is in m-ASL, horizontal distance is the distance from the ASR wells (in m)

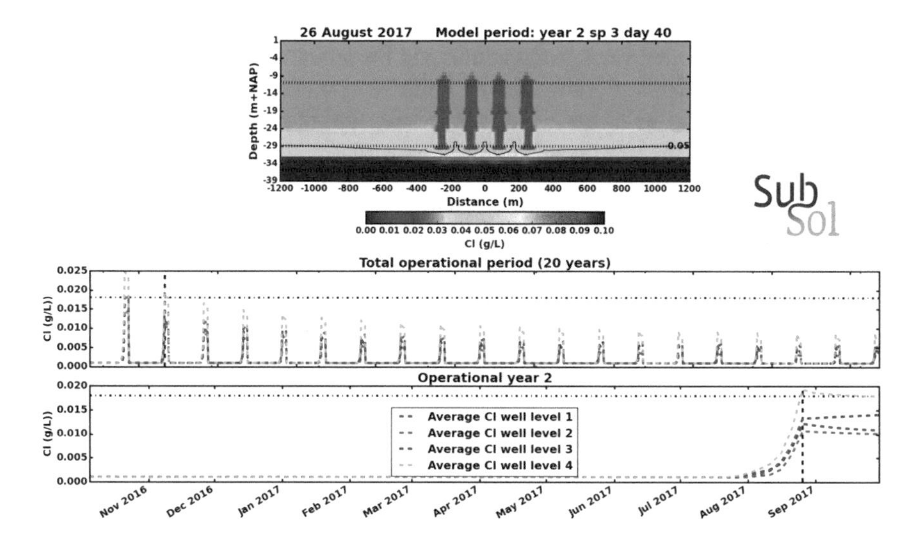

Fig. 8 The Dinteloord upscaled ASR-system: distribution of the injected water (upper), the recovered Cl-concentrations over 20 years (middle), and the recovered Cl-concentration in Year 2 (lower)

Table 2 Economics of the Dinteloord case

Storage type	Economic lifespan	Investment infrastructure	Reduced income by claim on land	Costs/ m^3
	years	×1000 euro	euro/yr	euro/ m^3
Reservoir	12.5	1875	3.6	3.09
ASR	20	780	3.6	0.46

reservoirs and claim on aboveground land. Most of the infrastructure for ASR can be used for at least 20 years (pipes, wells), while only pumps, valves, and sensors need regular replacement. Consequently, the price per m^3 supplied by the ASR scheme was much lower (Euro 0.46/m^3). On a yearly basis, the estimated resulted cost savings are more than Euro 300,000 per year.

The costs for the additional treatment (without storage) of the wastewater is Euro 1.50/m^3. The combination of irrigation water supply from rainwater (in individual basins, covering 80% of the demand) and the supplemented treated wastewater upon ASR results in an average cost price per m^3 of irrigation water of around Euro 0.60/m^3 for the total supply, which is in line with the Dutch price of high-quality irrigation water.

Many of the technologies discussed in the previous sections are currently implemented and commercially replicated. In most cases the setting is very specific, meaning that the technological, organizational and financial framing is tailored. MAR technology is typically not a one size fits all approach.

By now, the examples in the Netherlands of innovative MAR applications have spurred the dialogue between professionals and the authorities on how to make use of these local initiatives in view of future-oriented policies. In this case the challenge would not only be to match supply and demand of freshwater locally, but to create a framework for large scale applications with a synergetic effect on the scale of the western part of the Netherlands. Such an approach was recently developed, named COASTAR (COastal Aquifer STorage And Recharge). It aims at scaling up the ideas of MAR to a regional level. The vision is that private and public investments in MAR could very well function as building blocks of a broader, national strategic perspective to combat salinization and to balance future freshwater demand and supply.

However, there are two very practical obstacles that need to be overcome. The first one is the need for orchestrating the different components of such a strategy. In particular, to be effective on such a scale, spatial consistency in underground water management is an absolute prerequisite. Moreover, such an orchestration will take time. The same challenge goes for finding aggregated funding because short term private investments in horticulture not automatically match with the long-term perspective of a public sector seeking generic sustainability goals. At this very point in time one may even speak of a deadlock. On the one hand local private MAR projects are fundable because the underlying business cases are sound (as shown above) and the financial risks are low. However, privately funded projects are probably too small to be the sole fundaments and carriers for an effective and implementable strategy on a national scale. Large projects on the other hand may be too risky for investors, also because of the lack of ownership. This is one of the reasons why we believe that the financial engineering of regional and national MAR initiatives is a crucial success factor. Particularly, in case where private and public initiatives could create substantial synergetic effects. Therefore, financial framing of COASTAR should be critically looked at, considering both private and public investments. There are different models.

When the economic benefits of a project or program exceed the costs and the public sector has no specific role to fulfil, there is no reason for the public sector to be involved in financing projects. Most local MAR initiatives related to e.g. horticulture discussed here belong to this category (see also Table 2). When the short term economic benefits of a planned investment are lower than the costs, and a priced public good is at stake, in most cases the public sector is covering the integral costs by a system of taxes and/or tariffs for citizens. Financing the water infrastructure of the Dutch public water utilities is a good example. The Nederlandse Waterschapsbank (NWB Bank), a public National Promotional Bank (NPB), provides and arranges financial services in that context. NPB's are "Government owned financial institutions with the objective of fostering economic and social development by financing activities with high social returns" (United Nations 2005). In effect, NPB's could be considered as the national versions of international (regional) Development Banks. An example of the way they operate is a recent Water Bond (Green Bond) issued by NWB. It is used by its public shareholders, amongst others, the Dutch water authorities, for investments in adaptation measures in response to climate change. The costs are recovered because the resulting public services (water safety and security) are priced (tariffs, water tax etc.). COASTAR, until now in its aims and goals, is serving both private and public purposes. It provides technological solutions to private companies of which the costs are completely recovered through the price of their produce. However, their solutions (ASR) are way embedded in a regional hydrology that is not stable (salinization). The generic public functions of COASTAR should therefore be priced as well. After all, they support a generic environmental goal to be achieved and invites others to co-invest. Examples of such functions are combating salinization by ASR with treated wastewater and e.g. storing storm water belowground to prevent flooding after heavy rainfall. Not pricing such public goals means that potential benefits through synergetic effects with private investments (e.g. horticulture) are lost for the public cause.

The emergent instrument of Impact Investing addresses this kind of issue very adequately (De Nederlandsche Bank 2016; Glänzel and Scheuerle 2015; World Economic Forum 2013). It apportions value to both Economic (priced) and Social Benefits (not priced). The Social Benefits are based on the so-called ESG dimensions (Environmental, Social, and Governance) of a project or venture. Clark et al. (2015) conclude that "solid ESG practices result in better operational performance" and "investment returns." Furthermore, if this is relating to one's own business, or to the business in which one invests doesn't make much of a difference. Both are about material and immaterial rewards resulting from social responsible behaviour. These factors will probably and hopefully be the heart of the matter of any responsible business in the future. This way of thinking is also reflected in reports from many large Dutch pension funds such as APG (See their annual report, 2015; www.apg.nl/verantwoordbeleggen) and PGGM (Annual Responsible Investment Report, 2015).

In the case of the COASTAR, the aim is to look for innovative ways to price the generic environmental benefits. The environmental benefits related to storing much larger volumes of freshwater underground are substantial. One goal is to match water supply with demand better, another to prevent further salinization as well as flooding.

A water bank could possibly the right financial instrument to deal with pricing these public goods (Fig. 9). What it could establish is a functional connection between surface and groundwater management in the way Ghosh et al. (2014) describe that for the western USA. In their description of what a water bank specifically does, they see "merging surface water and groundwater rights into a single administrative framework" as a critical instrument for sustainable water management in water scarce areas. By performing that function a water bank is an economic instrument. It has been tested as a market instrument mainly in the USA, Australia, Chile and Spain (Megdal et al. 2014; Motilla-Lopéz et al. 2016). Most

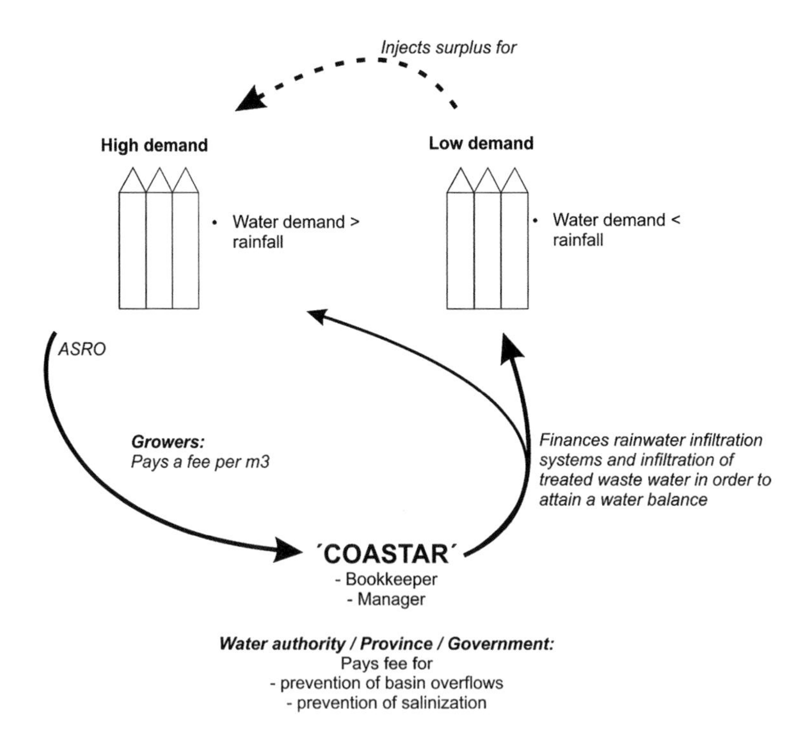

Fig. 9 The concept of a COASTAR water bank (Allied Waters 2016). 'ASRO' = aquifer storage and recovery and reverse osmosis'

cases have been developed under critical water scarce conditions. The instrument was originally defined as an efficient reallocation of existing water resources and specifically targeting "closed basins" (Motilla-Lopéz et al. 2016).

However, looking at the seasonal mismatch of freshwater supply and demand in the Netherlands, the instrument could work well under more temperate climatic conditions as well. For the COASTAR case it is particularly important to functionally link surface water and groundwater and to install and effective instrument for pricing both. That relates of course to the issue of ownership, assets or securities as collateral for loans and investments.

5 Conclusions

Wastewater use will increasingly gain importance as an alternative supply of water. The acceptance of the final users is important for the success of water reuse projects. The acceptability of reclaimed water is dependent on the physical, chemical, and microbiological quality of the water. A risk-based approach is needed for safe water reuse. Traditional indicator (*E. coli*) monitoring is insufficient since indicators are easily removed by treatment processes while pathogenic micro-organisms can be more persistent. Conventional wastewater treatment was not designed to remove pathogens, therefore post-treatment is needed for safe reuse. For MAR this can be limited to particle removal to prevent clogging.

Examples of successful implementation of MAR to catalyse safe and reliable water reuse are abundant. In the Netherlands, this started with the intake river water for dune infiltration in the 1950s, while the rivers were fed by upstream industrial and municipal wastewater discharges. To date, these big MAR schemes still function, supplying around one-fifth the drinking water in the Netherlands. Research has shown that these MAR systems are crucial for disinfection of the river water and overcoming mismatches between river water availability and water demand.

Finally, closing the water cycle by water reuse and MAR on a large scale with maximum benefits cannot exist without a coherent financing mechanism, involving all stakeholders that benefit from the integrated approach. This includes total cost recovery based on financing from all stakeholders organized in, for instance, a cooperation or water bank.

References

Allied Waters. (2016). COASTAR: A perspective for coastal freshwater management, www.alliedwaters.com.

Anonymous. (2001). Besluit van 9 januari 2001 tot wijziging van het waterleidingbesluit in verband met de richtlijn betreffende de kwaliteit van voor menselijke consumptie bestemd

water. (Adaptation of Dutch drinking water legislation) Staatsblad van het Koninkrijk der Nederlanden 31: 1–53.
Australian Government. (2008). *NWQMS: Australian guidelines for water recycling and augmentation of drinking water supplies*.
Clark, G. L., Feiner, A., & Viehs, M. (2015). From the Stockholder to the Stakeholder: How Sustainability Can Drive Financial Outperformance. Online:http://www.longfinance.net/programmes/london-accord/la-reports.html?view=report&id=464.
De Nederlandsche Bank. (2016). *Sustainable investment in the Dutch pension sector* (p. 33).
DESSIN. (2014). Demonstrate Ecosystem Services Enabling Innovation in the Water Sector. www.dessin-project.eu.
Dillon, P. (2005). Future management of aquifer recharge. *Hydrogeology Journal, 13*(1), 313–316.
Dillon, P., et al. (2006). Role of aquifer storage in water reuse. *Desalination, 188*(1–3), 123–134.
Dillon, P., et al. (2010). Managed aquifer recharge: Rediscovering nature as a leading edge technology. *Water Science and Technology, 62*(10), 2338–2345.
Ghosh, S., Cobourn, K. M., & Elbakidze, L. (2014). Water banking, conjunctive administration, and drought: The interaction of water markets and prior appropriation in southeastern Idaho. *Water Resources Research, 50,* 6927–6949. https://doi.org/10.1002/20144WR015572.
Glänzel, G., & Scheuerle, T. (2015). Social impact investing in Germany: Current impediments from investros' and social entrepeneurs'perspectives. *Voluntas*, https://doi.org/10.1007/s11266-015-9621-z.
Hornsta, L. M. (2013). Virusverwijdering door bodemtransport onder invloed suboxische condities. Nieuwegein, KWR: 93.
Megdal, S. B., Dillon, P., & Seasholes, K. (2014). Water banks: Using managed aquifer recharge to meet water policy objectives. *Water, 6,* 1500–1514. https://doi.org/10.3390/w6061500.
Montilla-Lopéz, N. M., Gutiérrez-Martin, C., & Gómez-Limón, J. A. (2016). Water banks: What have we learned from the international experience. *Water, 8*(10), 466. https://doi.org/10.3390/w8100466.
Olsthoorn, T. N. (1982). KIWA announcement 71: Clogging of injection wells (in Dutch), Keuringsinstituut voor waterartikelen, Niewegein.
Peters, J. H. (1995). Artificial recharge and water supply in the Netherland: State of the art and future trends. In: A. I. Johnson & R. D. G. Pyne (Eds.), *ISMAR 2. Proceedings of the Second ISMAR*.
Pyne, R. D. G. (2005). *Aquifer storage recovery: A guide to groundwater recharge through wells* (p. 608). Gainesville, Florida, USA: ASR Systems LLC.
Rook, J. J. (1976). Haloforms in drinking water. *Journal American Water Works Association, 68* (3), 168–172.
Schijven, J. F., Hoogenboezem, W., Nobel, P. J., Medema, G. J., & Stakelbeek, A. (1998). Reduction of FRNA-bacteriophages and faecal indicator bacteria by dune infiltration and estimation of sticking efficiencies. *Water Science and Technology, 38*(12), 127–131.
Smeets, P. W. M. H., Medema, G. J., & van Dijk, J. C. (2009). The Dutch secret: How to provide safe drinking water without chlorine in the Netherlands. *Drinking Water Engineering and Science, 2*(1), 1–14.
Stuyfzand, P. J., & Doomen, A. (2005). *The Dutch experience with MARS (Managed Aquifer Recharge and subsurface Storage): A review of facilities, techniques and tools* (Kiwa Report 05.001). Nieuwegein.
Tufenkji, N., & Elimelech, M. (2004). Correlation equation for predicting single-collector efficiency in physicochemical filtration in saturated porous media. *Environmental Science and Technology, 38*(2), 529–536.
United Nations. (2005). Rethinking the role of national development banks. Department of Economic and Social Affairs. Financing for Development Office. Background document 1. http://www.un.org/esa/ffd/msc/ndb/NDBs-DOCUMENT-REV-E-020606.pdf.
van der Wielen, P., Senden, W., & Medema, G. (2008). Removal of bacteriophages MS2 and ΦX174 during transport in a sandy anoxic aquifer. *Environmental Science and Technology, 42* (12), 4589–4594.

Vewin. (2014). Drinking water factsheet 2015.
Wetsteyn, F. (2005). Inspectierichtlijn Analyse microbiologische veiligheid drinkwater Artikelcode: 5318. VROM-inspectie. Haarlem, the Netherlands, VROM-inspectie.
World Economic Forum. (2013). From the margins to the mainstream. Assessment of the impact investment sector and opportunities to engage mainstream investors. https://iris.thegiin.org/research/from-the-margins-to-the-mainstream/summary.
Zuurbier, K. G., Raat, K. J., Paalman, M., Oosterhof, A. T., & Stuyfzand, P. J. (2017). How subsurface water technologies (SWT) can provide robust, effective, and cost-efficient solutions for freshwater management in coastal zones. *Water Resources Management, 31*(2), 671–687.
Zuurbier, K. G., & Stuyfzand, P. J. (2017). Consequences and mitigation of saltwater intrusion induced by short-circuiting during aquifer storage and recovery in a coastal subsurface. *Hydrology and Earth System Sciences, 21*(2), 1173–1188.
Zuurbier, K. G., Zaadnoordijk, W. J., & Stuyfzand, P. J. (2014). How multiple partially penetrating wells improve the freshwater recovery of coastal aquifer storage and recovery (ASR) systems: A field and modeling study. *Journal of Hydrology, 509*, 430–441.

Erratum to: Safe Use of Wastewater in Agriculture

Hiroshan Hettiarachchi and Reza Ardakanian

Erratum to:
H. Hettiarachchi and R. Ardakanian (eds.), *Safe Use of Wastewater in Agriculture*, https://doi.org/10.1007/978-3-319-74268-7

The original version of the book was inadvertently published without the credit line in the cover photo, which has been now included. The erratum book has been updated with the change.

The updated online version of this book can be found at https://doi.org/10.1007/978-3-319-74268-7

H. Hettiarachchi and R. Ardakanian (eds.), *Safe Use of Wastewater in Agriculture*, https://doi.org/10.1007/978-3-319-74268-7_9